Private Property of

Mrs Robt. Berksber

846 Grant St

Elkhart Ind.

46514

systems
engineering
applied to training

This series of books is supported by the American Society for Training and Development as part of its continuing program to encourage publication in the field. Most of the authors are active in ASTD and have contributed to its growth over the years. The Publications Committee of ASTD is a continuing link between the editor and the publisher and the membership.

systems
engineering
applied to training

leonard c. silvern

President
Education and Training Consultants Company

 Gulf Publishing Company
Book Division
Houston, Texas

Library of Congress Catalog Card Number: 72-76318
ISBN: 0-87201-814-8

To Elisabeth and Lisa

contents

foreword

It is doubtful if any single individual has contributed as much to the concept of systems in education as Leonard C. Silvern. It has been my personal pleasure to know him since the early 1950s and to have benefited from his insightful criticisms of some of my professional work. In the more than 20 years of our relationship, I have constantly found him about ten years ahead of the rest of us.

Today, hardly a book can be written in the HRD field without including something on systems. Yet, many of the authors are either trying out new ideas, or presenting us with "old wines in new goatskins." When Dr. Silvern uses the term systems, as he has for many years, he is quite specific about what he means and provides adequate materials so the reader can join him in pursuing the concept.

Within the Human Potential Series we have made room for varying ideas and concepts. We would not expect all our readers to agree with every book. We are still functioning in a field that contains varying parts of art and science. It is not necessary to belong to a particular "school." It is necessary to know the options available to those engaged in human resource development. In this context, we are proud of Dr. Silvern's addition to the series.

Leonard Nadler
Editor

preface

This is a book about training, but the training director, or better the "human resource developer," may not immediately recognize it as such. The illustrations do not resemble lesson plans or course outlines. The text doesn't present various methods of designing pre-test items. There isn't a description of how to use the chalk board or an explanation why an instructor should avoid jingling coins in his pocket while in class!

This book is written for training directors and personnel specialists who have long since passed through the *lesson plan-coin jingle* phase and realize that the key to success is organizing and managing properly. It approaches solutions systematically by using systems engineering techniques. One need not be an engineer to understand and apply engineering *concepts*. This text strives to interpret such concepts and describe applications in the typical training setting by using an elementary, step-by-step format. It attempts to avoid generalization yet is designed so that the concepts can easily be generalized to cover many different operational environments.

A way of thinking about human learning is presented here. Some of the concepts have evolved from Herbart's activities in the early 1800s, long before the training function in business, industry and government came into being. Other concepts were adopted from new-born industrial engineering at the end of World War I. The most significant contributions are from a part of electrical engineering known as *control systems*. In recent years, the term *cybernetics* has been given to the behavior of systems which interact with their environments. Also, technical concepts from the fields of information theory, coding, computing and from a growing body of knowledge called general system theory have been blended with earlier ideas to form the foundation for this book.

However, simply lifting concepts from a discipline like electrical and electronics engineering or chemical engineering and depositing them abruptly in the training environment can lead only to grief. Even a sophisticated human resources developer must *learn* this new language of systems engineering in order to apply the concepts in his work. There isn't anything mysterious about the techniques. All that is necessary is a desire to remain open-minded and acquisitive. Of course, open-mindedness has various definitions.

In preparing the manuscript, I have relied heavily on my experiences in training apprentices in the metal and wood trades; training technicians in electronics and weapon systems; training first-line supervisors; developing managers in public safety career fields; training instructors and systems designers; training instructional programmers; and similar activities which provide the realism so essential in systems design. Many of the bibliographic references are to earlier writings as well as the efforts of others. The field of systems engineering applied to training is so new that the number of worthwhile references is still quite small.

I have been influenced greatly by Carl N. Brooks, of Education and Training Consultants Company and now Manager of Advanced Systems in the Systems Research Laboratory of the Communications Division, Motorola Inc., particularly in quantifying what are essentially qualitative conceptualizations. Together, we continue to introduce order and logic into a system which has traditionally relied upon intuition and rules of thumb. There are many aerospace engineers and engineering psychologists who have contributed to and share the viewpoint expressed in this book. Finally, I am indebted to my former students at the University of Southern California, and at the University of California Extension, Los

Angeles, for challenging this position and implementing the ideas developed in the course. Special thanks to Dr. Leonard Nadler of the George Washington University, editor of this series, for encouraging me to share my views through preparation of this volume.

Information about innovation, by itself, will not produce innovation. Nonetheless, innovation is not likely to proceed without some information. This text hopes to provide sufficient energy to expedite the process.

Leonard C. Silvern
September 1971

systems engineering applied to training

1-systems

Every group of people communicates by using a language and those interested in systems represent such a group which must interchange ideas through language. Normally, we think of English or Spanish as "languages." Some languages are more symbolic than others—algebra is a more symbolic language than French. Of course, not all languages are understandable. We say that a confused, unintelligible language is a jargon. A jargon is also defined as the technical or secret vocabulary of a special group. Most special groups do have words, terms, phrases and symbols which are often unintelligible to outsiders.

If you wish to understand the meanings of the concepts expressed by a special group, you must learn its language. Once you do this, the communications are no longer gibberish. Systems people use symbolic languages such a flowchart diagrams and mathematical expressions. These are not really difficult to learn and use and in this book they are introduced gradually so you will have a high level of confidence as you proceed.

We say that a *system* is simply the structure or organization of an orderly whole, clearly showing the interrelations of the parts to each other and to the whole itself (1).

Figure 1-1. A rectangle symbolizes the whole system.

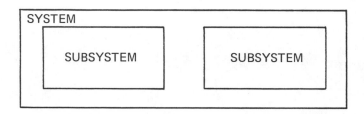

Figure 1-2. Subsystems collectively constitute a system and are embedded in it.

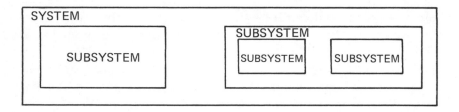

Figure 1-3. Subsystems embedded in a subsystem are known as subsystems.

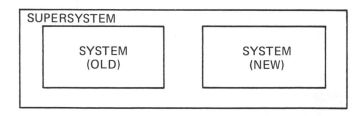

Figure 1-4. Combining an old system with a new system produces a super-system.

From this definition, these criteria emerge:

1. There must be a structure or organization.
2. The structure or organization must be conceptualized as a whole.
3. The whole must be orderly.
4. The whole must have parts.
5. Parts can be shown clearly relating to each other.
6. Parts can be shown clearly relating to the whole.

Parts are most often called *elements.* It doesn't matter if the part is an engine—or a spark plug—both are elements. When dealing with elements which are human actions rather than physical objects, the parts (elements) are usually called *functions.* The orderly whole is represented by a rectangle and a *descriptor,* consisting of words, which identifies the name of the whole. In Figure 1-1, the whole is the system.

If the system consists of two or more parts which interrelate, each part is known as a *subsystem.* The subsystems are said to be embedded in the system as shown in Figure 1-2.

When a subsystem consists of smaller elements, these are also *subsystems* as can be seen in Figure 1-3. The terms subsubsystem, subsubsubsystem, etc., are never used.

Until now, the system has been examined through the process of *analysis,* i.e., by breaking it down from a whole into parts. Assume that a system exists much like the lonely one in Figure 1-1. Suddenly, an element which was previously unrelated to this system is identified as having a clear relationship. Through *synthesis,* these are combined and a new and larger whole is created. Strictly speaking, the new element is a subsystem, but this would mean downgrading the old element from system to subsystem status. So, it is equally correct to retain system status for the old element by calling the new and larger whole a *supersystem* as in Figure 1-4.

One can conceive of a situation where a supersystem is operating and a remote, unrelated and different supersystem is found to relate and is then combined with the old one. The new and larger whole is the *suprasystem* depicted in Figure 1-5.

Finally, the combination of suprasystems into a new and larger whole results in the *metasystem* described in Figure 1-6. This is not quite as far out as it sounds if one considers multinational corporate structures, consortiums and conglomerates which are, in essence, the result of combining previously unrelated wholes.

Systems Engineering

Rau states that *systems engineering* consists of applying scientific methods in integrating the definition, design, planning, development, manufacture and evaluation of systems (2).

Chestnut selects a definition from among many, attributing it to Morton:

> The systems engineering method recognizes each system is an integrated whole even though composed of diverse, specialized structures and subfunctions. It further recognizes that any system has a number of objectives and that the balance between them may differ widely from system to system. The methods seek to optimize the overall system function according to the weighted objectives and to achieve maximum compatibility of its parts. (3)

Chestnut continues to examine the major precepts of systems engineering:

1. The idea of change.
2. Alternative ways of accomplishing goals.
3. Commonly accepted bases for judging the value:
   a. Performance
   b. Cost
   c. Time
   d. Reliability
   e. Maintainability
   Each system is in fact a subsystem of a larger system.
5. It may be better to produce a model and simulate before building and trying out the actual system.

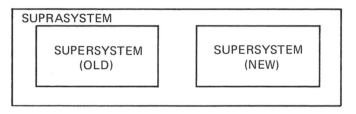

Figure 1-5. Combining an old supersystem with a new supersystem produces a suprasystem.

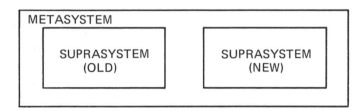

Figure 1-6. Combining an old suprasystem with a new suprasystem produces a metasystem.

The term *systems approach* has fallen into disrepute primarily because it has been interpreted in ways which weaken the original meaning (4). In this book, systems approach is a process consisting of four major parts: analysis, synthesis, modeling and simulation. These often follow in sequential order:

1. *Analysis* is performed on existing information to identify the problem, identify existing elements and identify the interrelation.
2. *Synthesis* is performed to combine unrelated elements and relationships into a new whole.
3. *Models* are constructed which can predict effectiveness without the actual implementation of the system.
4. *Simulation* is performed which reveals alternative solutions.

Silvern introduced the term *anasynthesis* for this specific process of analysis, synthesis, modeling and simulation (1). Anasynthesis is at least equivalent to the process inherent in the worn phrase systems approach and for many individuals is considerably more encompassing and the process more rigorous.

When anasynthesis is applied to training problems in a demanding manner, then anasynthesis and systems engineering are nearly synonyms. One might say that all anasynthesis is systems engineering—but all systems approaches are not anasynthesis.

Anasynthesis

The process of analysis, synthesis, modeling and simulation represents anasynthesis. In Figure 1-7, a flowchart model depicts the production of a system (1,6). The rectangles and descriptors are familiar, but numbers, arrows and (F) symbols have been introduced. The numbers are known as point-numeric code and appear in the lower right corner of each subsystem. The arrows are signal paths representing input to and output from a subsystem. A special kind of signal path incorporates the (F) signifying that it is a *feedback* signal returning to an earlier subsystem.

This is a simplified flowchart model incorporating the major precepts of change, alternatives, evaluation, modeling and simulation. A new system can be produced by following a specified sequence of events which are 1.0 to 8.0 in Figure 1-7. New ideas and new systems do not come out of the blue. Certainly this is true of training programs which require exquisite planning if they are to be successful. These new systems are first conceptualized in 1.0 of Figure 1-7. For the purpose of this discussion, a *training program* consists of courses, a course consists of units of instruction, a unit consists of lessons, a lesson consists of teaching points, and so forth (5). Thus, a training program or a course would constitute the system conceptualized in 1.0. The first subsystem (1.1) requires an analysis of existing or available programs, study of organizational needs, study of personnel needs, etc. Analysis (1.1) only deals with information that exists. As the result of analysis a management decision is made to create a new training program. The creation or synthesis occurs in 1.2 and usually consists of several solutions from which the best alternative is selected within 1.2.

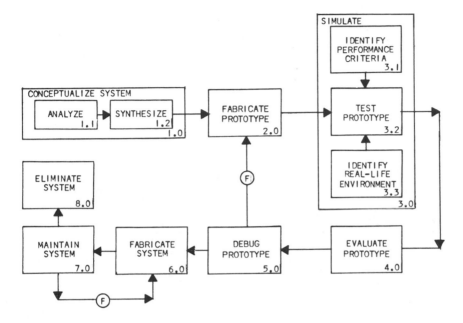

Figure 1-7. Model for producing a system.

The prototype is fabricated in 2.0. The term "prototype" is used rather than "model" to avoid confusion since Figure 1-7 is the model. A *prototype* is a test or pre-production version and in some industries is considered experimental. Fabrication is the making or manufacturing but not the "production" of the prototype since that term is usually reserved for the mass production function typified in 6.0. The training program prototype usually consists of software (curriculum materials) but may also include hardware (equipment, facilities, expendible supplies). Chestnut's major precepts call for model-building, which customarily refers to a paper model which can be simulated on a computer. However, it also includes prototypes as models which may be three-dimensional objects such as aircraft miniatures, and these can be placed in a simulated environment like a wind tunnel (1). In complex situations, there are paper models which go through paper simulation *before* an actual prototype is fabricated in 2.0 and simulated in 3.0. Complex training programs usually require this.

In 3.0 simulation occurs and the main objective is to test the prototype (3.2). However, a set of standards must be established earlier which will tell when the test is passed. These standards are *performance criteria* (3.1) which are applied to 3.2. Also, simulation should be conducted in a setting most like the actual setting in which the training is to be conducted. Ideally, the setting for training would also be where performance occurs after training has ended, but this one-to-one correspondence is rarely achieved in most training programs. The perceptive and experienced reader may comment that 3.1 is usually determined much earlier in the sequence, probably in 1.2, but is introduced in 3.1. The answer is that Figure 1-7 is a generalized model and does not attempt to explain every possible sequence of events—this will be presented later.

Real-life represents here-and-now reality and is portrayed by individuals who are "realists" or "pragmatists." In 3.3 it is important to identify the actual setting where performance occurs *during* training. For example, if a community (two-year) college level course in TV repair is to be instructed inside a penitentiary, this is delineated in 3.3, while 3.1 describes the performance criteria necessary for success in repairing TV sets *after release.* A prototype training program which has been designed in 2.0 for a nonreal-life environment, where 3.3 vanishes, is not very useful.

Following 3.2, an evaluation of the training program occurs in 4.0. This invariably turns up bugs or malfunctions. Very complex prototypes will have a large number of errors which must be eliminated.

Debugging of the prototype occurs in 5.0. Although the simple model in Figure 1-7 does not show it, evaluation (4.0) and debugging (5.0) rely on inputs from 3.1. Subsystem 5.0 has just two outputs. If debugging turns up no malfunctions, or very minor ones, the output is to 6.0. The more common situation is shown by a feedback signal path from 5.0 to 2.0 signifying that bugs have been detected and localized. It is essential to correct the prototype (2.0), retest it (3.2) and reevaluate it (4.0). Notice there is a clear image of a *closed loop* in the sequence 2.0 → 3.2 → 4.0 → 5.0 →

2.0. Feedback closes this loop and controls those events occurring within 2.0. Feedback signals *always* control an output. Modifying the prototype in 2.0 does *not* mean completely refabricating it. If testing in 3.2 is destructive, then 2.0 will call for a new prototype. With most training programs, testing is nondestructive.

Another reason for 2.0 modification is to eliminate bugs resulting from the process of altering design. Eliminating one bug can result in the introduction of other, different malfunctions. The closed loop 2.0 → 3.2 → 4.0 → 5.0 → 2.0 is also concerned with this aspect of alteration in which the cure may, on the whole, be worse than the disease.

When evaluation in 4.0 is accepted, the program passes through 5.0 into 6.0 where the production effort begins. There are essentially two levels of production: where training is instituted several times and is terminated; and where it continues over and over again because of a very large number of trainees. In the former case, training offered only several times may be of great importance. For example, training the Apollo Flight Controller team for a relatively short time has produced significant success in the NASA program (7). In general, 6.0 deals with producing training programs which are conducted through a number of cycles. Also, 6.0 represents production of training programs which are to be conducted on an identical basis at many different locations by decentralized staffs.

Maintenance in 7.0 means the operation of the program, which includes all activities necessary for it to continue generating graduates who satisfy the performance criteria of 3.1. The system may be maintained for one, five or even twenty-five years. In the fantastic societal change of the 1970s, it is unrealistic to stipulate long life for any training program. Products and procedures change and training for product support must keep up. Services and procedures change and training must be provided. And the mobility of the labor force in the United States suggests that training is necessary for those representing minority groups and others being transferred. Social systems in the modern world outlive their usefulness. They must change and coexist with the dynamic society or supersystem of which they are parts. A training program is, of course, one element of a social system and should be designed to be maintained for two, five, eight or ten years—at most.

During operation and maintenance (7.0), bugs are found as the result of experience. These bugs are corrected in ongoing programs within 7.0. However, if a series of identical programs are being produced in 6.0, it is necessary to feed back these corrections from 7.0 to 6.0 so new systems will be improved. The feedback from 7.0 to 6.0 is a feedback signal path which produces an error-correcting closed loop. Feedback signals *always* control an output, in this instance the output of 6.0. Such feedback tunes up 6.0 and usually does not output to 2.0, which would call for a major change. Theoretically, it is possible to have feedback from 7.0 to 2.0, but the model in Figure 1-7 does not show it. These feedback loops provide quality control, but they could be used for cost control as well.

Systems which have outlived their usefulness should be eliminated, as in 8.0. This might be sudden death or graceful retirement. If managment tends to avoid a decision eliminating a program, then it should be planned during design for the program to self-destruct at a future time. Programs which run forever are usually disharmonious with the environment in which they are embedded, and the entire supersystem is in disequilibrium. Feedback is not the sole technique to maintain equilibrium—self-destruction *by design* is another. Feedback and self-destruction by design are engineering terms for "working within the establishment" or "working within the system" used by young people and activists today. The alternative is to abolish a system from outside, which is anarchy or chaos.

Figure 1-7 represents a model for producing a system in which the events of analysis (1.1), synthesis (1.2), modeling (2.0) and simulation (3.0) represent anasynthesis. This model is closed-loop in that two feedback loops exist to control the process. Shouldn't a feedback signal output from 8.0 input to 1.1? Despite the short life of present-day training programs, those persons who began 1.1 are rarely on the scene when 8.0 has been achieved. This is particularly true in business, industry and government, where training is an interim assignment in a continuum of other activities. Because of this, and also because people take great pleasure in reinventing the wheel, an output from 8.0 to 1.1 when a *new* system is being conceptualized is an unrealistic expectation. A designer developing a new system in 1.1 concurrently as the old system in 8.0 is being phased out should analyze data in 1.1, which includes an input from 8.0 revealing experience with the system being retired.

Cybernetic Systems

Having examined the flowchart model for the production of a system, one may ask if the concept in Figure 1-7 represents a cybernetic system.

Porter states that cybernetics is concerned with the communication and manipulation of information and its use in controlling the behavior of biological, physical and chemical systems. Its implications are being applied in economic and social planning (8).

DeGreene thinks of cybernetics as an attempt to understand organisms through analogies to machines, understanding a given process in terms of information content and flow in terms of feedback and control (9).

Klir and Valach say that cybernetics is a science dealing, on the one hand, with the study of relatively closed systems from the viewpoint of their interchange of information with their environment, on the other hand with the study of the structures of these systems from the viewpoint of the information interchange between their elements (10).

Silvern defined *cybernetic model* as "an equation, a physical device, a narrative, or a graphic analog representing a relatively closed system revealing the interchange of information with its environment as well as the structure and organization of the system revealing information interchange between its internal elements" (11). He further described an *instructional system* as cybernetic when it "is an information processing system; all of the functions within an instructional system deal with information processing. . . there is feedback (information interchange) which provides stability and equilibrium. . . the system is closed–loop. . . "(12).

In essence, the central concept in cybernetics is information. The control aspect in cybernetics is provided by feedback signal paths. The model in Figure 1-7 *is* a cybernetic model and a training program developed according to the model would be a cybernetic system. However, it is possible that pressures of company policy, time shortages, desires to save money, etc., can distort this model. While most subsystems would be functioning, some would be inoperative. The program would be less successful in the short run and probably unsuccessful on a long-term basis. All subsystems are critical and all signal paths are essential, but some are more critical and essential than others. From a cybernetics viewpoint, 3.1 and 3.3 deal with information of the environment to which graduates flow and these are supercritical. The feedback path from 5.0 to 2.0 is superessential because it controls 2.0 and prevents inappropriate programs from being produced in 6.0. The feedback signal from 5.0 to 6.0 is superessential since it controls program quality in upcoming programs. 8.0 is supercritical because it eliminates programs unsuited to the operating environment.

LOGOS—A Language for Modeling Systems

It should be obvious that a graphic technique such as drawing a flowchart model furnishes the author of a concept with a clear and simple method of communication. The reader as a receiver of the communication can understand nearly everything in the mind of the author as transmitter. There may be a fallacy here. The author must put down on paper, in flowchart form, exactly what is in his mind. If he fails to do this, communication is fuzzy. A well-defined language will force the author to select his words or other symbols carefully and this requires his explicit and precise understanding of what he wishes to say. Also, both author and reader must use the same language so that the message is unambiguous.

A *language* is a set of representations, conventions and rules used to convey information. If these rules are poorly described, then communications will be weak. The specific language to be described here is LOGOS, an acronym derived from the expression *L*anguage for *O*ptimizing *G*raphically *O*rdered *S*ystems (5, 13, 14). The term *logos* itself is a derivation of the Greek "word or form which expresses a thought." In philosophy, logos represents the rational principle in the universe. All of the elementary LOGOS rules necessary to read and draw flowchart models are explained in this section. Since they are all together, locating a particular rule later should present no problem.

LOGOS is a language which communicates the thought or concept embodied in a group of words or string of characters of another language. LOGOS is a graphic language; i.e., it does not rely solely on alpha (a, b, c ... z) and/or numeric (1, 2, 3 ... n) characters or groups of words, but utilizes other shapes and symbols representing the fundamental vocabulary elements.

Elementary applications of LOGOS rely on alpha characters forming groups of words or narratives which are combined with LOGOS symbols culminating in a flowchart. Advanced applications depend upon mathematical equations which supplant groups of words and which are combined with LOGOS symbols to culminate in a flowchart. Thus LOGOS is used to communicate effectively with those readers preferring words and also with others who insist on the unambiguous terminology of mathematics. LOGOS is a language used in model-building. The thought expressed by a LOGOS flowchart is a conceptualization in the form of a graphic analog representing a real-life situation. In this frame of reference, a LOGOS flowchart is a flowchart model.

Function

A *function* (known also as functional block or subsystem) is represented by a rectangle, a descriptor and a point-numeric code as shown in Figure 1-8.

```
┌────────────────────────────┐
│           IDENTIFY          │
│            PRIOR            │
│          TRAINING           │
│                     2.2.2   │
└────────────────────────────┘
```

Figure 1-8. A function is represented by a rectangle, descriptor and code.

1. *Rectangle*: a right-angled parallelogram proportioned approximately 5:9 and oriented so that the longer side is parallel with the top and bottom of the paper, chart or other display medium.

2. *Descriptor*: a group of one to five words consisting of numeric and/or alpha characters in natural language, which collectively express the function or real-life usefulness of an object, action or information precisely and unambiguously. Objects are usually identified with nouns. Actions are usually identified with an imperative or action verb; present participles are less frequently used. In Figure 1-8, the action verb is "identify." Information is expressed by a group of one to five words which communicate most effectively. The descriptor is centered inside the rectangle if there are no subsystems. In a rectangle with two or more subsystems, the descriptor is located at the upper left corner. Descriptors are printed in upper-case alpha characters.

3. *Code*: the point-numeric code is always located in the lower right of the rectangle. In Figure 1-8, the code is 2.2.2. The major functions are coded 1.0, 2.0, 3.0. . . n. Together they constitute the first level of detail and always depict the total system (or supersystem, suprasystem, metasystem). Logically, one rectangle would depict a total system coded 0.0, but this is infrequently required unless there is reason to believe that a future synthesis will identify, relate and combine 0.0 with one or more undefined systems. At that point, 0.0 becomes 1.0, and appropriate changes in code occur throughout the flowchart model.

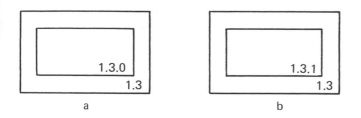

a   b

Figure 1-9. Illegal uses of coding for only one subsystem embedded in a larger one.

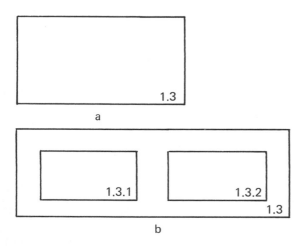

a

b

Figure 1-10. Legal uses of coding for a single sybsystem (a) and two subsystems embedded in a larger one (b).

It is *not* permissible for one function to consist of just one other function and be coded as in Figure 1-9 *a* or *b*.

A function may consist of itself or of two or more parts described in Figure 1-10*a* and *b*.

Conceptually, the *level of detail*, is the level of specificity or resolution of a particular system. For example, Figure 1-10*a* is fairly general in comparison with Figure 1-10*b*. Figure 1-10*a* is at the second level of detail while 1-10*b* is at the third level of detail.

Table 1-1

Determining Level of Detail

Code Structure	Reasoning	Level of Detail
1.0	Has one position	First
1.3	Has two positions	Second
1.3.1	Has three positions	Third

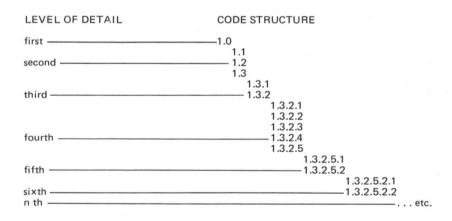

LEVEL OF DETAIL   CODE STRUCTURE

first ——————————————————————1.0
                                          1.1
second —————————————————— 1.2
                                          1.3
                                              1.3.1
third ———————————————————— 1.3.2
                                                  1.3.2.1
                                                  1.3.2.2
                                                  1.3.2.3
fourth ———————————————————— 1.3.2.4
                                                  1.3.2.5
                                                      1.3.2.5.1
fifth ——————————————————————— 1.3.2.5.2
                                                          1.3.2.5.2.1
sixth ——————————————————————— 1.3.2.5.2.2
n th ————————————————————————— . . . etc.

Figure 1-11. Level of detail correlated with code structure.

Mechanically, in terms of point-numeric code, the level of detail is determined by counting numeric positions (see Table 1-1).

The code structure and levels of detail are correlated in Figure 1-11.

When a model has only major functions coded 1.0, 2.0, 3.0... n, it is said to be at the first or major level of detail. This is

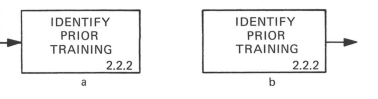

Figure 1-12. Input (a) and output (b) signal paths.

an important psychological factor when a training specialist attempts to explain a complex flowchart model to his management. Unless he first uses a model at the major level of detail to introduce the general concept, use of a detailed model usually results in poor communication.

Signal Path

Signal path (known also as path or arrow) is represented by a straight, solid line having an arrowhead at one end. The signal path carries information or data in the direction shown by the arrowhead. The complexity of information carried depends on the level of detail. One signal path symbolizes 1, 2, 3 ... n discrete streams of information. A specified signal path may carry information only in one direction.

1. *Input signal path.* Information entering a function is represented. The arrowhead is solid and touches any side of the rectangle. The line is usually perpendicular to the side and is shown in Figure 1-12a.
2. *Output signal path.* Information exiting a function is represented by Figure 1-12b. The tail of the signal path is at the line termination opposite the arrowhead. The line is usually perpendicular to the side. (One signal path usually represents an output and an input as in Figure 1-13.) Notice that an effort is made to place the action verb alone on the first line of descriptor text.

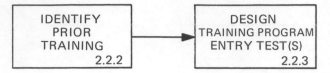

Figure 1-13. Information flow is symbolized by an output from 2.2.2 and an input to 2.2.3.

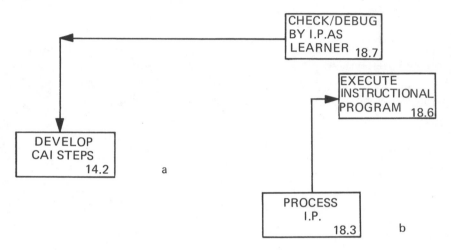

Figure 1-14. Arrowheads at 90° turns are required in a but not in b.

3. *Redirected signal path.* A 90° change in direction requires one arrowhead at the turn, as in Figure 1-14*a*. However, very short redirected paths do not conform to this requirement, as in 1-14*b*. If the paths are very long, one or more arrowheads are permitted in the direction of signal flow.

4. *Input to all functions within a subsystem.* When *one* output signal path is input to *all* functions within a subsystem, either *a* or *b* in Figure 1-15 is permitted.

5. *Output from all functions within a subsystem.* When *all* functions within a subsystem are output to *one* function, either *a*

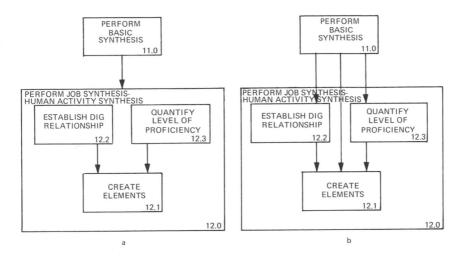

Figure 1-15. Alternative methods a and b for describing input to several embedded subsystems.

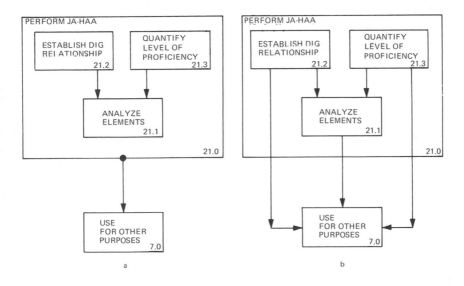

Figure 1-16. Alternate methods a and b for describing output from several embedded subsystems.

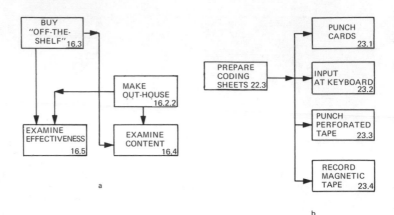

Figure 1-17. Symbolism for crossing signal paths.

Figure 1-18. Input and output relating to distant functions.

or *b* in Figure 1-16 is permitted. The diameter of the tail is
equal to the width of the arrowhead.

6. *Crossing signal paths.* When one signal path crosses another
   and there is *no* relationship, the representation is *a* of Figure
   1-17. When one signal path crosses another and there is a
   relationship, one arrowhead at the intersection establishes the
   relationship *b*.

7. *Input or output from distant functions.* When a signal path
   relating two functions is not major and is extremely long, the
   technique in Figure 1-18 is permitted. However, this technique
   is *not* recommended for feedback signal paths and should be
   avoided.

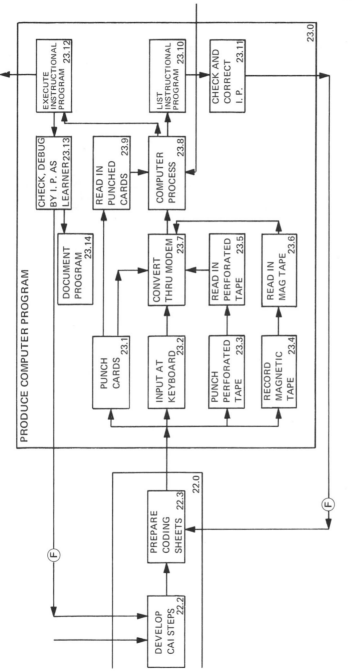

Figure 1-19. Feedback is represented by F *in the signal path.*

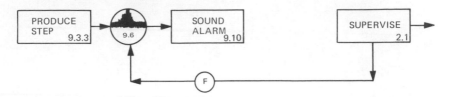

Figure 1-20. Feedback to an error signal function.

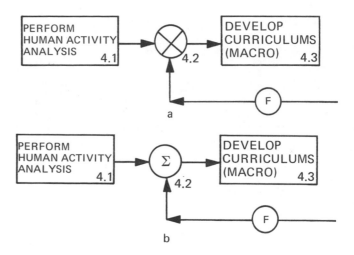

Figure 1-21. Feedback to summer functions.

Feedback Signal Path

Information may be output from a subsystem and input to any preceding subsystem. If the signal path creates a closed-loop and controls the output(s) of the same preceding subsystem, it is a *feedback* signal path. Feedback signals always control the output of a function. The F symbol is located on the flowchart so that it may later be supplanted by a rectangle containing a mathematical equation or expression.

Feedback may input the preceding subsystem in one of several ways. The most common appears in Figure 1-19 and is recommended for training activities. A second representation is as input

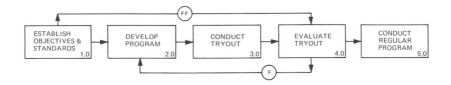

Figure 1-22. Feedforward is used to provide information to a succeeding subsystem. (Feedback controls a preceding subsystem.)

to an *error signal* function depicted as a circle containing an oscilloscope display in Figure 1-20, used when graphic emphasis is required. A third is as input to a *summer* or *summation* function depicted as a circle containing crossed lines in Figure 1-21*a*. The summer function always has an output which controls the output of the function succeeding the summer. An alternative method appears in *b*. These are used primarily by electronics engineers, human factors engineers and systems engineers but are quite suitable for training programs where quantification is demonstrated.

Feedforward Signal Path

Information may be output from a subsystem and input to a succeeding subsystem. If there are intervening subsystems unaffected by this specific signal path, it is a *feedforward* signal path as in Figure 1-22. The (FF) symbol is located on the flowchart so that it may later be supplanted by a rectangle containing a mathematical equation or expression. The feedforward symbol is limited to models at the first level of detail. The information in the signal path may *not* control the function which it inputs and is unlike feedback in that characteristic.

Special Symbolism

1. *Signal paths converging to a dot.* When two or more subsystems relate to each other but the precise relationship is unknown, signal paths converging to a dot are used for represen-

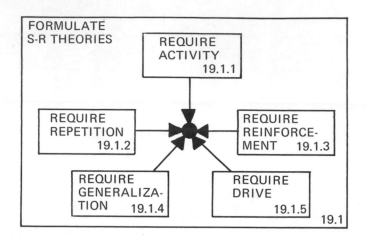

Figure 1-23. Imprecise relations are described by signal path convergence to a dot.

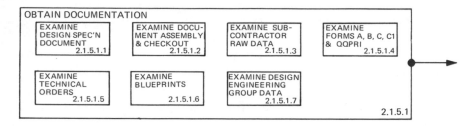

Figure 1-24. Clustering elements which appear to group.

tation as in Figure 1-23. Contrast this with Figure 1-24. In some instances, it is not possible to show even an imprecise relationship. The subsystems *appear* to fit together but evidence is lacking. The symbolism in Figure 1-24 is used with a recommendation that greater attention be given to synthesis before the model is finalized.

2. *Alarm.* Input to alarm is always a feedback signal. It symbolically causes a light to illuminate (as in the balloon over a cartoon character) or a bell to ring (as inside the ear) alerting the subsystem to the error, as in Figure 1-25.

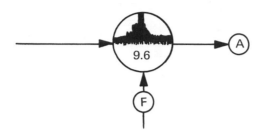

Figure 1-25. Alarm symbol used to focus attention on feedback and error signal.

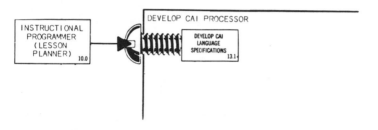

Figure 1-26. The screw symbolizes an input from without which can modify CAI language specifications developed within 13.1.

3. *Machine screw*. Adjustment by human to a *critical* function is depicted symbolically by a screw (Figure 1-26). The input signal path enters the head and outputs at interface between end of screw and subsystem rectangle. A critical function is one which is so essential that the system will fail if it malfunctions. In most cases, the screw has a feedback input since feedback always controls an output and the screw symbolizes adjustment or change. Obviously, LOGOS deals with dynamic systems and *all* the subsystems are subject to influences resulting from input variations. The screw merely identifies critical signal inputs and differentiates them from other signal inputs.

4. *Limit switch and alarm*. Adjustment by human to a critical function may go beyond the design limits for that function. The limit switch in Figure 1-27 symbolically detects an input which exceeds design tolerances. If the screw is turned in too far, the head is sheared off; turned out too far, it drops off and

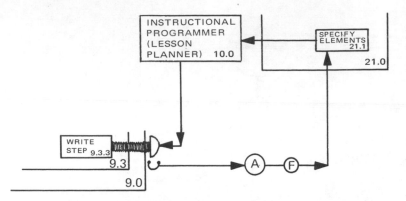

Figure 1-27. Screw travel symbolizes design tolerance. The limits activate an alarm representing feedback.

is lost on the floor! It activates an alarm, and the feedback signal inputs a preceding subsystem to control its output. The reader is cautioned to approach the limit switch concept and similar designs with care. Flowchart models must be simple if they are not to become "Rube Goldbergs" (5, 15). The late Rube Goldberg said to design engineers in 1969, "The equipment and systems you design are too complex. . . extra components increase the probability of failure. . . how many control systems work sporadically because some engineer wanted to add a very clever, complex circuit when a simpler one would do?. . . designing simple circuitry isn't simple. . . a good engineer. . . should look back to see what he has designed *in* that can be designed *out*."

5. *Filter function.* A filter is used to suppress, enhance, eliminate or otherwise modify a signal or parts of a signal. There are many filter designs, but those of immediate interest deal with separating a signal from noise which accompanies the signal.

6. *Noise.* Any spurious or undesired disturbances that tend to obscure or modify the desired signal are generally defined as noise (3). Noise increases the uncertainty of communication and reduces the probability that a message will be understood. In Figure 1-28*a* there is a *distortion filter* to which a distorted

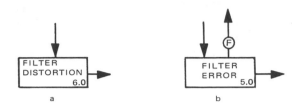

Figure 1-28. The simple filter in a *has one output. The filter in* b *has two outputs of which one is feedback.*

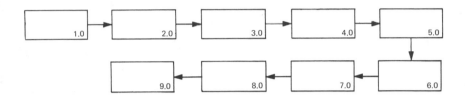

Figure 1-29. Locating and positioning subsystems when paper is too small.

signal inputs. The filter (6.0) removes any distortion permitting the "purified" signal to output. The distortion is internally attenuated much like heat loss in an electric wire and later dissipated. The *error filter* in Figure 1-28*b* differs since there is no internal attenuation in 5.0. Instead, that part of the signal carrying error information is fed back to a preceding subsystem and it controls an output as is customary with all feedback signals.

Sequencing and Coding

1. *Position and location.* In open-loop models the first subsystem is coded 1.0 and is positioned on the left side of the paper, chart or other display medium. The succeeding subsystems are sequenced from left to right. If the paper is not sufficiently wide, the flow should output down and across as in Figure 1-29. In closed-loop models, the first subsystem is coded 1.0

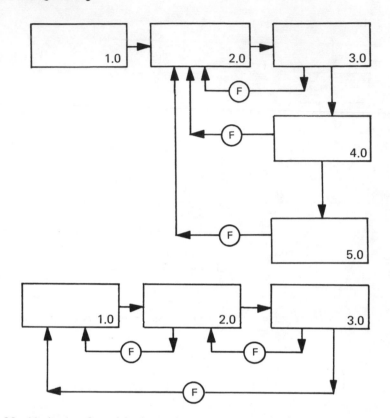

Figure 1-30. Methods of positioning subsystems to emphasize closed-loop concepts at the major level of detail.

but may be positioned at any location, preferably on the left side of the paper. It is important, particularly at the major level of detail, to show feedback as a clearly distinguishable closed loop as in Figure 1-7. Other ways of representing closed loops clearly appear in Figure 1-30.

2. *Coding.* Normally, codes are assigned sequentially to functions which are naturally or logically ordered. Where the ordering is not natural or logical, the codes are assigned arbitrarily. Codes are like license plates and merely identify elements in terms of levels of detail.

Figure 1-31. An open-loop flowchart model of a sequence of events in baseball.

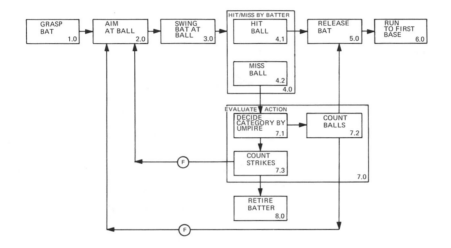

Figure 1-32. Performance evaluation and alternative events with feedback producing a closed loop.

3. *Routing.* Routing of signal paths and positioning of functions within a subsystem are arbitrary and the designer's choice. However, the rules above apply to all levels of detail within every element.

Open-Loop and Closed-Loop Concepts

To understand the importance of the closed-loop concept, one must first be familiar with the open loop. A simple six-subsystem configuration appears in Figure 1-31 using LOGOS language to depict a sequence of events in baseball.

The model is very specific. No "strikes" are allowed. No "balls" are allowed. One pitch is made apparently by the pitcher and in 4.0 it is hit and the player moves toward first base (6.0). There is no evidence that first base is reached successfully. There is no provision in this particular model for any alternate events—such as missing the ball in 4.0. Figure 1-31 implies two players: the batter and the pitcher. However, a third person judges the events, and in Figure 1-32 the umpire function is shown. The batter now has *alternatives* in 4.0. He can either hit (4.1) or miss (4.2). Notice the use of the input signal permitting entry to either subsystem. A hit signifies the sequence 4.1 → 5.0 → 6.0. If he misses (4.2), an official decision is necessary in 7.1 and counters summarize in 7.2 or 7.3. This describes reasonable alternatives; the model does not cover *all* judgments by the umpire.

If the strike count is within limits 7.3, then a feedback signal to 2.0 controls the output of 2.0 which, in reality, controls the aiming and swinging behaviors of the batter. The feedback signal from 7.2 to 2.0 also controls these behaviors, but the control resulting from balls produces a strategy different than from strikes. The model allows a batter to be retired as the result of strikes 7.3 → 8.0. It also allows him to run to first base 7.2 → 5.0 → 6.0. The model does *not* consider other events at or near first base which may cause an "out" or similar situations.

The closed-loop 2.0 → 3.0 → 4.2 → 7.1 → 7.3 → 2.0 controls the output of 2.0 and also imposes quantification on the loop. The strike counter (7.3) is reset to zero when event 1.0 occurs. When it records three strikes, the next event is 8.0. In the same manner, the ball counter (7.2) records up to four balls then 5.0 → 6.0 occurs. During the action, the closed loop is 2.0 → 3.0 → 4.2 → 7.1 → 7.2 → 2.0.

A training program is analagous to the events in Figure 1-31. The training program events in Figure 1-33 were taken from a foods corporation formula used in the 1940s (16). One becomes aware that the model does not incorporate a change mechanism, nor does it authorize alternative solutions or methods. It is, in fact, almost identical with the open-loop baseball model which assumes that the ball will be hit! Suppose that training (2.0) cannot correct employee inadequacies (1.0). Perhaps certain em-

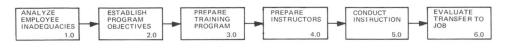

Figure 1-33. An open-loop flowchart model of a sequence of events in training.

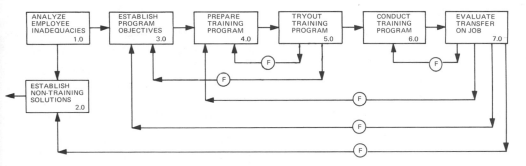

Figure 1-34. Performance evaluation (7.0), alternative events (2.0) and feedback loops in a closed-loop training model.

ployees are not trainable for the behaviors found to be essential. Suppose the objectives are clear (2.0) but ordinary methods of instruction in 3.0 are unable to produce the terminal behaviors. Where is the small-scale tryout during 5.0 which permits the training director to debug the program and run it again before releasing it to full-scale operation? How does evaluation in the plant change 2.0, 3.0 or 5.0 to optimize the system? A closed-loop version of the same project is shown in Figure 1-34. If employee inadequacies are corrected by nontraining solutions (transfer, termination, medical treatment, etc.), the flow is 1.0 → 2.0. A tryout phase (5.0) allows the program and its instructors to make two kinds of corrections before proceeding to 6.0. Program preparation may require error correction 5.0 → 4.0 or, if the outcomes are seriously off target, reconsideration of objectives (3.0) is necessary. Such changes would follow the path 5.0 → 3.0 → 4.0 → 5.0.

The greater challenge, however, is represented by the four feedback signal paths from 7.0, for it is in this subsystem that practical use of the knowledge, skills and attitudes is made. If the training program should be modified so that corrections are of a relatively minor nature, feedback is to 6.0. Changes can be inserted into the next program without going through tryout. If corrections are major, feedback from 7.0 to 4.0 requires that redevelopment occurs and tryout (5.0) follows. If the job (7.0) has changed materially, then feedback is to 3.0, suggesting that objectives be reestablished and a new program prepared (4.0). Another output feeds back from 7.0 to 2.0 indicating that training, per se, is *not* the solution to employee inadequacies and a nontraining solution should be investigated and established (2.0).

Open-loop systems tend to be single- or few-solution networks. If the environment is dynamic, the system is not designed to adapt to change. Open-loop systems work well when the events are totally predictable and each event occurs in the proper sequence and on time. These systems are suited for one-man training organizations where control, if any, is within a single person who performs all tasks. This includes design (2.0) in Figure 1-33, development (3.0), instruction (5.0) and evaluation (6.0). However, when supervision and delegation are involved (and this occurs in nearly every training program in the United States) then the typical open-loop system is not good.

Closed-loop systems tend to incorporate alternative solutions and have various branches representing the "if-then" logic. Modeling and simulation occur in several stages. In Figure 1-34, a paper model is created in 3.0 and orally simulated. A higher level of modeling occurs in 4.0 when the program is prepared and it is simulated in 5.0, usually on a sampling of trainees with the same characteristics as the target population. The model in 3.0 is a flowchart model. The model in 4.0 is a full-scale replica of the training program. In this sense, 4.0 corresponds in Figure 1-7 to *fabricate prototype* (2.0), and tryout in 5.0 represents *simulate* (3.0) of Figure 1-7.

Finally, one must understand that all closed-loop systems do not have a feedback signal path. Figure 1-35 represents a closed-

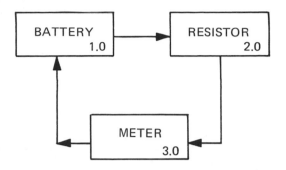

Figure 1-35. Closed-loop system without feedback.

loop system. 1.0 stores and produces current flow. This current inputs (2.0) and there is an output. This output is also the input to 3.0, and the meter produces a reading or readout if it is digital. Current leaves 3.0 and inputs 1.0. So long as the system remains closed, current will flow and the meter will provide a reading. Losses eventually will cause the battery to lose its storage of energy and the reading of 3.0 will ultimately decay to zero. If feedback is considered in the conventional sense of controlling an output, adaptive control or quality control, there is none in Figure 1-35, yet it has a closed-loop.

At times, the flowchart models in this book will have signal paths which may *suggest* that they might be feedback paths. The Ⓕ signifies feedback. The logical question always must be, *Does the signal return to a preceding subsystem and control an output?* If it goes to a subsequent function, *it is not feedback.* If it goes to a different subsystem within the *same* functional block, the question of precedence still applies, Does it return to a preceding subsystem and control an output? Figure 1-36 presents signal paths which are certainly identified as feedback paths but could be in error. The path 4.0 → 2.0 is quite correct; 2.0 precedes 4.0. However, 4.0 is *not* related sequentially to 3.0 except that the paths *appear* to be parallel on a time continuum. Remember, the point-numeric codes are not always reliable indicators of sequence but identify elements as to levels of detail. The 4.0 → 3.0 path

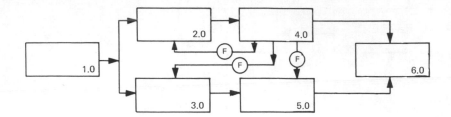

Figure 1-36. Feedback and non-feedback in a system.

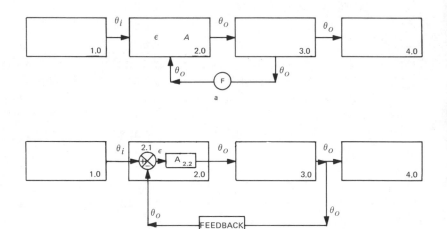

Figure 1-37. a. *LOGOS depiction of feedback;* b. *engineering representation of feedback.*

may or may not be feedback. Logic dictates that it is *not*, since the output it controls (3.0) does not influence 4.0—it is not a preceding subsystem. The same logic applied to $4.0 \to 5.0$ suggests that it, too, is not feedback.

This brings the discussion to the technical characteristics of feedback using algebra and, with it, the first chapter comes to an end.

Closed-Loop Configuration Utilizing Feedback

Feedback controls an output. The LOGOS analog in Figure 1-37*a* states that a signal from 3.0 controls the output of 2.0. The engineering equivalent is shown in Figure 1-37*b*. The output of 1.0 is θ_i and it is the input to the summation function (2.1). Now, jump downstream to 3.0. The output of 3.0 is θ_O. A feedback of θ_O returns to 2.1. Thus the relationship between the two inputs is shown by the equation

$$\epsilon = \theta_i - \theta_O$$

where ϵ is the error signal.

The output of (3.0) is proportional to the error signal ϵ and is related to it by the constant in 2.2, which is A. A represents the open-loop *gain* of 2.2. The equation is

$$\theta_O = A\epsilon$$

If the feedback loop is *disconnected*, then 2.1 sees θ_O as zero and the error signal ϵ is equal to the input signal θ_i. When θ_O is 0, then

$$\epsilon = \theta_i - 0$$

or

$$\epsilon = \theta_i$$

This is the *open-loop* condition equation reasoning:

$$\epsilon = \theta_i \qquad \text{(when open loop)}$$

$$A = \frac{\theta_O}{\epsilon} \qquad \text{(substituting for } \epsilon)$$

$$A = \frac{\theta_O}{\theta_i}$$

It can be seen that the open-loop gain A of a closed-loop system is the ratio of output θ_O to error ϵ. Under *ideal* conditions, the gain A should be unity. This means that the ratio of θ_O to θ_i should be one or unity.

In designing a subsystem, the effort is made to make the open-loop gain high so the closed-loop gain is approximately unity. Here is the reasoning:

$$\epsilon = \frac{\theta_O}{A}$$

$$\epsilon = \theta_i - \theta_O$$

Equating

$$\frac{\theta_O}{A} = \theta_i - \theta_O.$$

Multiplying both sides of equation by A

$$\theta_O = A\theta_i - A\theta_O$$

Moving terms

$$\theta_O + A\theta_O = A\theta_i$$

Factoring

$$\theta_O(1 + A) = A\theta_i$$

Transposing $(1 + A)$

$$\theta_O = \frac{A\theta_i}{1 + A} = \frac{A}{1 + A}\theta_i$$

Transposing

$$\frac{\theta_O}{\theta_i} = \frac{A}{1 + A} \tag{1-1}$$

Thus the final equation in this series represents the mathematical relationship in Figure 1-37 between the feedback input θ_O to 2.1 and the regular signal input θ_i to 2.1 in the *closed-loop* condition.

To illustrate, the gain of 2.2 is $A = 9$. Therefore, using equation 1-1:

$$\frac{\theta_O}{\theta_i} = \frac{A}{1 + A} = \frac{9}{1 + 9} = \frac{9}{10} = 0.9$$

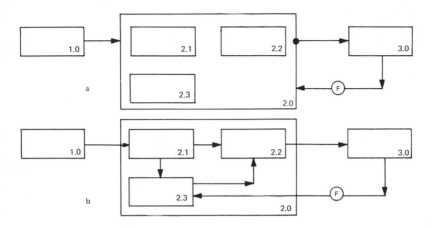

Figure 1-38. a. *Generalization of feedback and interrelationships;* b *specificity of signal flow, sequence and feedback control.*

meaning that the feedback θ_O has an effect on the input of 0.9 due to the gain of the subsystem (2.2). If the gain A is increased to 99, then

$$\frac{\theta_O}{\theta_i} = \frac{A}{1+A} = \frac{99}{1+99} = \frac{99}{100} = 0.99,$$

meaning the feedback θ_O has an effect on the input of 0.99, but it can never exceed 1.0 or unity.

The general statement of equation 1-1 is:

$$\frac{\text{output}}{\text{input}} = \frac{\text{open-loop gain}}{\text{open-loop gain} + 1}.$$

Now, this concept may be applied first to Figure 1-37a and suggests that feedback θ_O *controls* the output of 2.0 by mathematically controlling the error signal ϵ. In electrical systems, the signal θ_O is relatively simple. However, in social systems, of which training programs are a subset, the feedback θ_O can be more complex.

The method of reducing complex feedback to a simpler form is to break down the elements to a level of detail which avoids generalizations and draw feedback signal paths from one specific element to another so there are no ambiguities. In Figure 1-38, *a* represents generalization and ambiguity while *b* depicts specificity and clarity.

In this text, the mathematical treatments are used only occasionally to illustrate the quantification techniques available to the training director if he is able to go beyond the flowchart modeling and simulation emphasized. It is beyond the scope of this text to mathematize flowcharts. The reader interested in such techniques is invited to read materials of an advanced nature (2, 3, 10, 17, 18, 19).

2-a model for improving human performance

Essential Relationships

In this chapter, a flowchart model is developed which attempts to identify *all* of the essential elements in the improvement of human performance through training. While the model is learner-centered, emphasis will be on its curriculum effectiveness aspects. It is basically a *management* planning and process model with foundations going back many years (1, 21). To a degree, it was influenced by Forrester's work in industrial dynamics but it has a cybernetic flavor which may be considered unique (22). Certainly the invention of LOGOS makes it possible for the average training specialist to analyze elements and their relationships without the mathematics and computer simulation of industrial dynamics.

The relationship described in Figure 2-1*a* deals with the *consumer* (2.0) and the *producer* (1.0). The reader, if he is a training specialist, should consider himself in the producer role. In *b*, the action verbs *consume* and *produce* conform to LOGOS standards and yield more accurate descriptions. Products and/or services leave 1.0 and enter 2.0 where they are processed. A stock broker (1.0) buys a stock at the request of 2.0 and delivers the certificate.

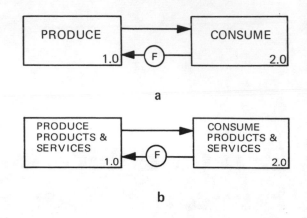

a

b

Figure 2-1. Essential relationships between producer and consumer.

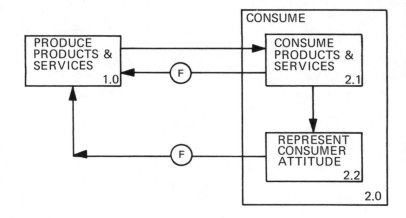

Figure 2-2. Feedback mediated through 2.2.

A forest ranger (1.0) delivers a campfire talk which is received by visitors (2.0) at a national park. An airline pilot (1.0) flies a passenger (2.0) to Chicago. A waitress (1.0) takes an order and serves a meal to a customer (2.0). The list is endless and incorporates *all* relationships between those who produce (1.0) and others who consume (2.0). In these examples, it is crystal clear who the *consumer* is—the consumer is the customer who pays for the product or service. Remuneration is not shown as an element in Figure 2-1

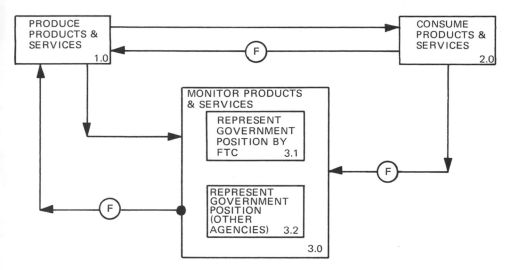

Figure 2-3. Feedback from 2.0 and mediated through 3.0.

and is not necessarily a condition in the relationship. The feedback paths represent information fed back to the producer to *control his product or service.* Feedback signals always control an output and, in Figure 2-1, the output is understood to be the *quality* of the product or service. In some instances, feedback controls *cost* or pricing. When it affects both, feedback controls the *cost-effectiveness* relationship with respect to the consumer.

Mediated Relationships

In recent years, the $1.0 \leftrightarrows 2.0$ relationship has been strained by an apparent disinterest of 1.0 in the feedback content. The clearest example of a solution is shown in Figure 2-2 in which a consumer-oriented group, represented by an organization such as Nader's Raiders, feeds back data to 1.0 with the express purpose of controlling the *quality* of 1.0 output. This model suggests that individual consumer feedback from $2.1 \rightarrow 1.0$ is unable to control 1.0 output and an independent organization (2.2) with a crusading attitude represents consumer input $2.1 \rightarrow 2.2$ and feeds this back $2.2 \rightarrow 1.0$. Figure 2-3 describes the Federal Trade Commission and other federal agencies which are able to mediate feedback received from consumers (2.0) as complaints and also to create feedback as the result of independent monitoring within 3.0. In both instances, feedback controls output. Because the individual consumer cannot adequately represent himself and thereby control the output of 1.0 to 2.0, he must rely on a government watchdog agency such as FTC (3.1). In Figure 2-3, either FTC (3.1) or another government agency (3.2) processes the input to 1.0 where just three states can exist:

1. 1.0 can ignore feedback from 3.0 which attempts to control it.
2. 1.0 can be controlled: $3.0 \rightarrow 1.0$.
3. 1.0 can negotiate some type of settlement: $3.0 \rightarrow 1.0 \rightarrow 3.0 \rightarrow 1.0$, which finally has some effect on the output of 1.0 to 2.0.

The mediated relationships modeled in Figures 2-2 and 2-3 represent functions which are *outside* the producer subsystem and organizationally not a part of it. Figure 2-4 describes a model in which 1.2 is a function to mediate external feedback. This receive and analyze function (1.2) is within and organizationally a part of the produce subsystem (1.0). Historically, the 1.2 function was titled "customer service" or a similar name with the purposes of accepting consumer input, performing minor tasks to placate the customer, explaining producer policy and procedures, and generally to act as a barrier between consumer views (2.0) and producer policy in generating products and services. If a product was very poorly designed, or even unsafe for the user, 1.2 attenuated such feedback signals from 2.0. One need not bring to mind automobiles, cigarettes, contaminated foods and improperly manufactured pharmaceuticals as examples of products which affect consumer health.

Governmental functions are not exclusively confined to *monitoring* as in 3.2 of Figure 2-3. A "Consumer Advocate has been appointed by the Postmaster General . . . he will head a newly created Office of Consumer Affairs reporting directly to the Postmaster General. . . suggestions or complaints. . . should be ad-

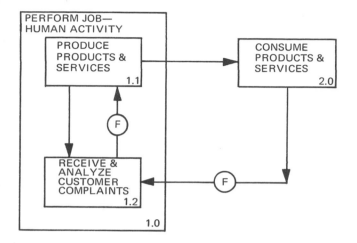

Figure 2-4. Feedback to a specific function (1.2) within the produce subsystem (1.0).

dressed. . . "(20). This consumer advocate is the complaint receiver (1.2) in Figure 2-4. As an avid user of the U.S. Postal Service, the author hopes that the feedback will not be absorbed and signals lost in the Office of Consumer Affairs! Feedback *always* controls an output. However, there are different degrees of control. A housewife buys a table through mail order and finds it is delivered "knocked down"; i.e., she is required to assemble it. She finds:

1. An incomprehensible set of instructions.
2. One bolt is missing.
3. A wing nut breaks as it is tightened.

In feeding back to 1.2 in Figure 2-4, assuming that management has set up ideal internal channels to improve the product line, the signal paths would appear like those detailed in Figure 2-5. 2.0 would alert the company (1.3) by complaining. Upon receipt of this complaint 1.3 would alert the marketing function (1.2). Hopefully the missing bolt (2.1.3 → 1.0), the broken wing nut (2.1.3 →

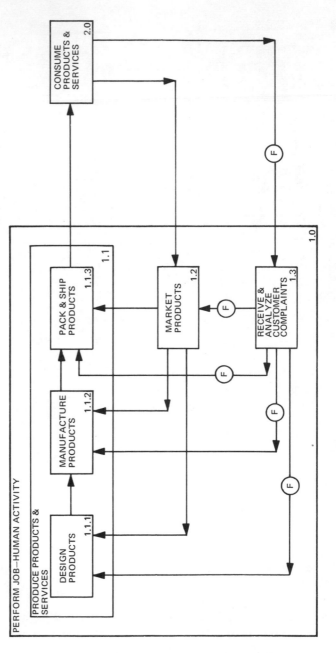

Figure 2-5. Four internal feedback signals resulting from one unhappy consumer (1.0).

1.0) and individual assistance in assembling the table would reach the consumer (1.2 → 1.1.3 → 2.0). These are quick-fixes which might pacify one consumer but really do not solve the basic problem—something is wrong inside the 1.1 function.

Missing bolts in knocked-down products are intolerable since the entire sale is predicated on consumer assembly. Thus, either more intensive supervision or an industrial engineering solution to improve inspection, packaging or both is indicated in 1.1.3. Training might be a solution but the nature of the missing bolt problem points to foolproofing the procedure rather than the employee. The feedback is 1.3 → 1.1.3 to *control* hardware packaging.

Wing nuts which break off during assembly are also intolerable. Unhappy customers—even lawsuits—could result. Feedback 1.3 → 1.1.1 would call for a design review. Why are wing nuts used? Are the material specifications correct? Are they too small? Feedback 1.3 → 1.1.2 would raise the question of quality control by the manufacturer and supplier of wing nuts. Training is not the solution. Purchasing controls and design philosophies are technical management functions which are not cured by a training program for lower level employees.

Poorly written instructions reveal insensitivity to the intellectual acuity of an average consumer. Feedback 1.3 → 1.1.2 should set into action a sequence of events which might begin with training—training the instruction sheet writer to break steps down, to use simple illustrations, to use close-up sketches, etc.

The mail order table problem illustrates a low-level application of feedback from consumer to producer. Suppose the product can produce death or disfigurement even if used *properly*. Cars which flip over and cigarettes which cause cancer are convenient examples. Then, the feedback is essentially directed to the *design* function. In Figure 2-6, a synthesis of mediating functions depicts a present-day model relating producer, consumer and monitor. This model may appear to be incongruous in a text on training, but the main reason for training employees is to produce products and services

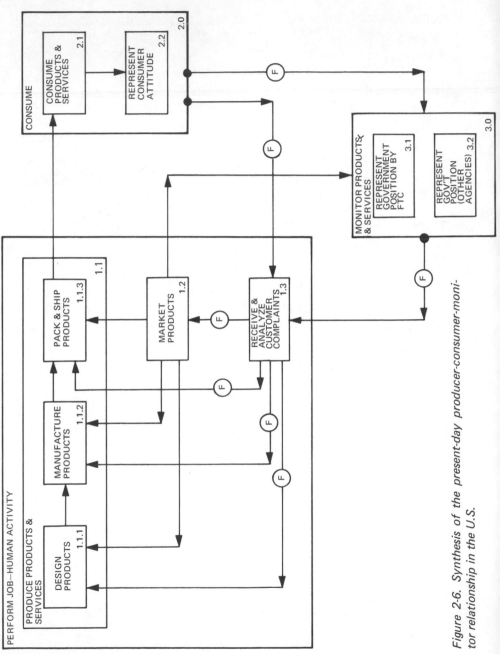

Figure 2-6. Synthesis of the present-day producer-consumer-monitor relationship in the U.S.

which are accepted by the ultimate consumer. Otherwise, there would be no need to train.

Major Relationships within the Producer Organization

Training is a peculiar kind of activity barely tolerated in a business or industrial organization despite hard evidence that many ills can be solved by carefully designed training programs. It is also true that badly designed courses have induced malaise into an already feverish environment. While training is a *way of life* in a military establishment because both manpower and technology are in a constant state of flux, it is often a *way of death* in other organizations. Training per se cannot *produce* employment, for example; it can only prepare individuals to be successful if a job is available.

The key to success in training is *organization*—the logical, meticulous design and implementation of a structure which is orderly. This organization can be manifested as a flowchart model, but it can never be described through the use of an organization chart.

This model should be cybernetic and have these characteristics:

1. Graphic analog, specifically a flowchart model.
2. Relatively closed system revealing the interchange of information with its environment.
3. Revealing the interchange of information between its internal elements.
4. All functions should process information.
5. Feedback provides stability and equilibrium and is based on closed loops.
6. Processing complex problems yields alternative solutions.

We have already reviewed models relating a consumer and producer; Figure 2-6 is a realistic analog of information interchanges and feedback signal paths which generally describes several closed loops. Producing a model for a training system should follow the

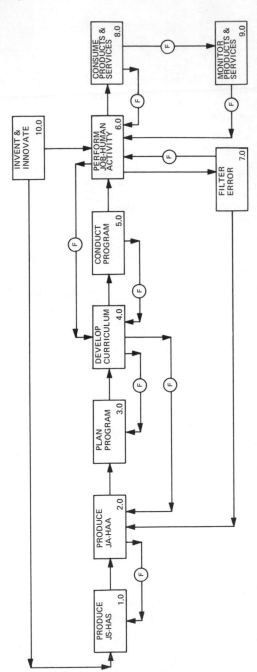

Figure 2-7. Ten subsystems constitute the "Training System" which includes performance (6.0), consumption (8.0) and monitoring (9.0) in the real-life environment.

procedure elucidated in Figure 1-7 for producing a model for any system. Therefore, the *general* steps are:

1.0 Conceptualize system
2.0 Fabricate prototype
3.0 Simulate to test
4.0 Evaluate prototype
5.0 Debug prototype
6.0 Fabricate system
7.0 Maintain system
8.0 Eliminate system

The *specific* steps are:

1.0 Produce job synthesis-human activity synthesis
2.0 Produce job analysis-human activity analysis
3.0 Plan program
4.0 Develop curriculum
5.0 Conduct instruction
6.0 Perform job-human activity
7.0 Filter error
8.0 Consume products and services
9.0 Monitor products and services
10.0 Invent and innovate

These appear in Figure 2-7.

Conceptualization of the program usually begins in 2.0 where analysis of existing employee behaviors occurs. "Behaviors" in this sense are the various skills, knowledge and attitudes essential for performing activities in 6.0 which are acceptable by company supervision. As a result of these behaviors, the consumer in 8.0 receives products and services. "Behavior" in the disciplinary sense has a different meaning and will not apply to this discussion. There are a number of methods for examining employee current behaviors ranging from direct observation, including analyst experience, to questionnaires. All of them follow this sequence: 6.0 → 7.0 → 2.0. The only reliable data source is the real-life environment of the job where the day-by-day action is. It is in the job environment that first-line supervision controls and evaluates employee performance and where training program graduates are assigned to produce products and services. During this process, the analyst who is highly skilled may run across procedures which may be illogical to him. In questioning the supervisor, the analyst may show that the procedures are erroneous and should be returned as feedback from the filter error function 7.0 to 6.0 where they control *two* outputs: $6.0 \rightarrow 7.0 \rightarrow 2.0$ and $6.0 \rightarrow 8.0$.

The first designs are produced in 3.0 when such system parameters as *number of graduates, levels of proficiency required, time to train, budget support* and *staff/facilities commitment* are considered. These are management parameters which are independent of 2.0 but follow sequentially.

Fabrication of the prototype occurs in 4.0 where the curriculum is produced and materials are prepared. *Simulation* to test the model occurs at a conceptual level. *Simulation* to try out the curriculum is performed at an operational level in 4.0. Regardless of the type of instructional method, the curriculum does not move to 5.0 until one or more cycles of simulation at the operational level are satisfactorily completed. At this time, errors in planning are corrected by feedback path $4.0 \rightarrow 3.0$, thereby controlling the output of 3.0. Errors in the job analysis-human activity analysis (known as JA-HAA), consisting usually of incomplete or missing facts, are corrected by a feedback path $4.0 \rightarrow 2.0$, thus controlling the output of 2.0. Each iteration is followed by *evaluation* and *debugging* until the training specialist is willing to accept the performance which the program will produce. At that moment, the program leaves development (4.0) as a fully *fabricated product* and enters 5.0 where it is conducted.

To *maintain* means to operationalize or conduct on a full scale and to provide whatever logistics are necessary to keep the program functioning according to specifications. This occurs in 5.0 and may be months or years long depending on system parameters in 3.0. As the individual programs end, new programs begin. Each time, improvements are fed back from 5.0 to 4.0 as new and better ways to instruct are found. However, these are usually microlevel modifications—not global changes. Those who complete a program in 5.0 have an opportunity in 6.0 to transfer what has been learned to real life in the shop, office, store, factory or wherever they work to produce products and services.

Transferring what has been learned to real life occurs in 6.0 and is measured as output from that function in the form of increased volume of products or services or increased quality. While difficult to attribute to a specific employee in a large company, changes in output are measurable as "lumped" values. The output becomes an input to 8.0 and, under average conditions, the system is in one state of equilibrium and there are no feedback inputs. Not only will the consumer not complain but he won't increase consumption either! Another state of equilibrium exists when the consumer (8.0) complains (8.0 → 6.0) and there is product improvement which results in happier consumers who then increase their consumption. It should be clear that training per se may not be the reason why customers are happy and increase their input. One glance at Figure 2-6 will testify that marketing, which includes advertising, quality control in the manufacturing process and product design, has greater influence on consumer attitudes than training. Of course, designers, manufacturing personnel and marketers must be trained and thus we can justify the need for trained personnel as the need for training.

About 99% of newly planned training programs deal with occupations which have existed for many years. Therefore, the most logical path in Figure 2-7 would be 6.0 → 7.0 → 2.0 → 3.0 → 4.0 → 5.0 → 6.0 → 8.0. It is also true that job content is dynamic and does change, but supervision in 6.0 is responsible for keeping updated, and it can be assumed that the job, as it exists, will be the best resource for job content. Where job practices are simply poor and supervision is not rotated or replaced, new methods would be introduced through the vehicle of the training program. The analyst in 2.0 would do more than merely analyze existing job practices—he would take on responsibilities of the industrial engineer. This tends toward synthesis rather than analysis and may violate company policy in large organizations. In theory, the analyst should not be expected to create new job procedures or set job standards of employee performance.

Less than 1% of programs under consideration at any time involve occupations or tasks which have never existed before. The world is admittedly dynamic and new products and services are constantly being generated inside that mystical black box in Figure 2-7 called *invent & innovate* (10.0). However, most of this is innovation, where the existing occupation is *modified*, rather than invention where an entirely new occupation is created. The most outstanding example of invention in recent years is in the computing industry, where new tasks seem to be created almost daily. A dying occupation is typified by the railroad industry, where conductors, brakemen, firemen and engineers are employed. Science-based organizations tend to require more job synthesis-human activity synthesis (1.0) because their products and services rely on invention and innovation to a considerable extent.

This brings up the matter of *lead-time*. When the United States was young, there was virtually no research in business and industry; nearly all research was performed in universities (23). Concentration was on design. Some work was done on development but, because products and services were relatively simple by present-day standards, the main emphasis before World War II was on manufacture. The factory was nearly, if not quite, the most important part of a company whose aim was to produce. An employee was introduced to the job he was to perform *almost at the instant he was to perform it.* It was unthinkable in those days to train the employee before placing him in his job. Between the two world wars virtually no organized, formalized training existed. This period has been aptly termed "the standstill years" (24). Employees were not trained and it was expected they would acquire the necessary skills and knowledge without any assistance from management. During the latter part of World War II the United States faced the problem of preparing an unbelievable number of men and women civilian employees as well as armed forces members for an almost infinite variety of jobs, mostly of a technical nature.

As new and more complex jobs were formulated, particularly because of the use of electronics, a new training philosophy began to evolve. Employees had to be trained *before* the jobs were actually ready for them. Had there been a delay until the system to be operated was physically available, there would have been a further delay before it became operational.

It was in the national interest during the war to place both civilian and military equipment in service and have it immediately productive. An increasing need arose for the personnel who were to operate this equipment, whatever it was, to be trained in advance. After the war, research became prominent in business, industry and government, and testing of products and services was the rule rather than the exception. The systems introduced were increasingly complex. After 1946, employee training started at an earlier stage of manufacture—long before the product was to be delivered. This training sometimes began one, two or even three years before the system itself was to be put in operation. In fairly complex systems, the employees may be trained for the system when it is in the design or development stage, and in a number of instances three to four years may elapse between the time employee training starts and the time the system is put into operation. The trend generally seems to be towards more complex systems, which suggests that employees may be introduced to training earlier. It is even conceivable that this introduction could occur during the final stages of research. Certainly the NASA Apollo project falls into the category of training astronauts and associated personnel during research, in preparation for an operational application years ahead. Lead-time increases as job content grows more complex.

In essence, all invention and innovation in the universe with which we deal occurs in subsystem 10.0 of Figure 2-7. If such contributions are significant, they will gradually enter the operational situation in 6.0, expand and become a way of life. If insignificant, they will reside in a few 6.0 organizations and ultimately disappear. A large organization must rely on an early-warning surveillance function to alert itself that a particular object, procedure or bit of information is vital. This becomes an input to 1.0 from 10.0 and signals a series of events where early information on research or preliminary design begins to generate job information. As products become more complex, the job content grows more complex and lead-time increases. Once job synthesis is achieved (1.0), it is possible to review it and, as time goes on, to analyze it in 2.0, feeding back 2.0 → 1.0 until the job content is thoroughly understood. This feedback loop exercises control over the quality of 1.0 output.

Now we must consider the *eliminate system* (8.0) subsystem shown in Figure 1-7 to see how this can be incorporated into Figure 2-7. As has been said, systems which have outlived their usefulness should be eliminated. First-line supervision, if it is carefully selected and managed, should be in the best position to identify changes in their departments. With some assistance from training specialists, they should be able to eliminate a particular training program. Many programs continue almost indefinitely due to internal politics, making jobs for older employees, etc., but the most common reason can be pinned down to ignoring the real needs of the supervisor whose people are being trained. If the supervisor is paying for training from his budgeted funds, he can simply turn the spigot and training ends. Quite often, training is centrally budgeted and the supervisor is limited to turning just the manpower valve! He merely refuses to supply a trainee on the grounds that the employee cannot be spared. Clearly, this is an abuse of supervisory authority but it is an old trick commonly practiced. The best way to terminate a training program is for the training specialist to detect the need for elimination and act to shut down the program. However, this may be suicidal in an empire-building world since no self-respecting individual would dissolve a program with himself a part of it. Elimination decisions might better be made by a higher level of management.

Other Relationships within the Producer Organization

Having examined ten subsystems which constitute the closed-loop, cybernetic system depicted by the model in Figure 2-7, we are ready to expand it through synthesis to include these new functions:

11.0 Perform basic synthesis
12.0 Perform basic analysis
13.0 Establish selection criteria
14.0 Formulate learning psychology
15.0 Provide human-instruction
16.0 Provide machine-instruction
17.0 Apply human performance engineering
18.0 Operate logistical support
19.0 Provide facilities
20.0 Develop CAI processor
21.0 User's group improve CAI processor
22.0 Develop CAI courses
23.0 Produce computer program

This synthesis increases the original model from 10 to 23 subsystems and, in so doing, provides a rather complete analog of real life. Later, each of these 23 will be further analyzed to reveal the detailed structure of the elements and signal paths.

Figure 2-8 represents the 23-subsystem model which also incorporates Figure 2-7. The term *job synthesis* (1.0) is synonymous with human activity synthesis, and *job analysis* (2.0) is the same as human activity analysis. While there is no apprehension over the adjective "job" in business and industry, specialists in education avoid the terms *job analysis* and *job synthesis* preferring to use *task analysis, task synthesis, behavioral analysis, operational analysis,* etc. The terms *human activity synthesis* and *human activity analysis* were created basically to satisfy demands of academicians who dislike associating human activity with vocational activity. "Job" meant vocational education to them.

Silvern defines job as a "unit of work accomplished by a worker which carries him through the entire unit from beginning to end" (25). He defined job analysis as an "intensive, direct method for obtaining job facts relating to items of work performed;

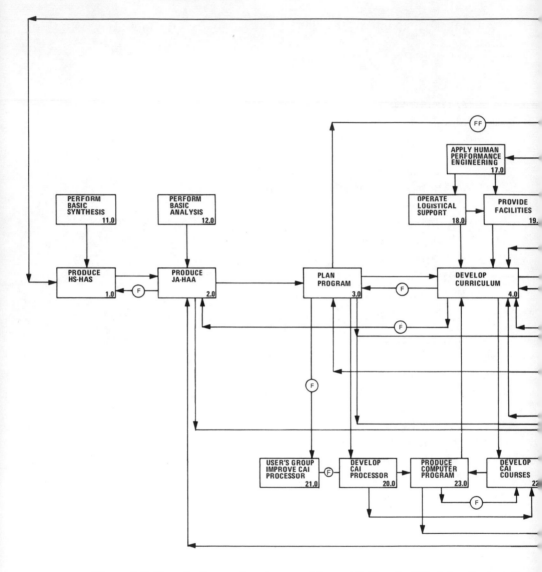

Figure 2-8. Twenty-three subsystem model constituting the "Training System."

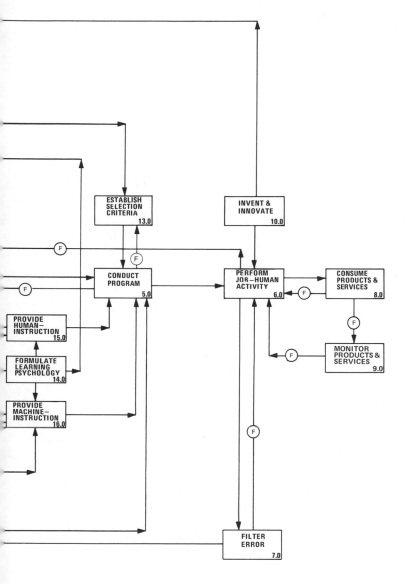

tools, equipment and materials used or worked on; skills, knowledge and physical demands associated with the *job*" (25).

The academician will not recognize that a mother performs a job and a child performs a job just like a milk deliveryman performs a job. In the same manner, one can conduct a job analysis of mother, child and milk deliverman with equal ease. To bridge the semantic gap, "human activity" was created, but in this book the adjective job will continue to be used.

A job synthesis is used when new jobs are created in real life and training is to be developed *before* the jobs are fully formed. Thus, training is synthesized concurrently with job synthesis. This calls for a person with synthesis capability who performs basic synthesis (11.0) regularly to produce the job synthesis in 1.0; the relationship 11.0 → 1.0 reveals the direction of information flow. Synthesis (11.0) consists of steps involving *identifying, relating, combining* and *limiting:*

1. Identify a bit of information.
2. Identify a different bit of information.
3. Relate these parts to each other.
4. If there is a relationship, combine the parts to form a new whole.
5. Limit by halting the process.

Later, a model of the process in 11.0 will be described in greater detail.

Analysis is quite different from synthesis. While synthesis combines essentially unknown and unrelated parts into new wholes, analysis is the process of relating known parts and separating them into smaller parts while maintaining the relationship of parts-to-wholes. Analysis (12.0) consists of steps involving identifying, relating, separating and limiting:

1. Identify the whole piece of information.
2. Identify the parts of the whole.
3. Relate the parts to each other.
4. Relate the parts to the whole.

5. Separate the parts.
6. Limit by halting the process.

Later, a model of the process in 12.0 will be presented. The information flow is 12.0 → 2.0, signifying that job analysis (2.0) relies on basic analysis (12.0) to be performed.

Some mention has been made of the program planning function (3.0), Figure 2-8. One set of decisions in 3.0 has to do with the makeup of the trainee population. Such parameters as *title, level, paygrade, longevity of service, prior experience, age, immediacy of training* and others are considered in establishing selection criteria in 13.0. The (FF) or feedforward signal path 3.0 → 13.0 indicates decisions made in 3.0 are operationalized in 13.0 and the path 3.0 → 4.0 signifies other selection decisions are implemented in 4.0. Actual trainee selection, based on 13.0 criteria, assignment through scheduling, occur in 5.0, shown by 13.0 → 5.0.

Aside from the content, another set of decisions in 3.0, Figure 2-8, has to do with the *method* of instruction. If the program is to be instructed in the conventional sense, or developed so communications is under direct instructor control, the decision 3.0 → 15.0 is to provide human-instruction. The instruction is provided to 4.0 for tryout purposes and also to 5.0 for the full-scale, operationalized activity. Suppose the program is to be in nonhuman-instruction format? It could be a film, TV presentation, sound filmstrip, programmed instruction text, teaching machine program, training simulator, etc. Such a decision (3.0 → 16.0) is to provide machine-instruction. Generally speaking, machine-instruction does not require the mediation of an instructor and thus differs from human-instruction. Those methods which are truly "programmed instruction" are characterized by these factors:

1. Participative, overt interaction or two-way communication between learner and machine or program.
2. Lesson sequence is carefully controlled and consistent.
3. Learner receives immediate knowledge of his progress (input) which is a feedback signal to control his subsequent behavior (output) and reinforce it.

4. Learner learns at his own rate.
5. Machine mechanism in sophisticated devices shapes and controls learner behavior.

The path $16.0 \to 4.0$ represents the machine-instruction decision which calls for program tryout and validation in 4.0. The path $16.0 \to 5.0$ depicts operationalizing the validated product in the full-scale activity (5.0). The requirement to try out and validate may apply only to those materials *actually created* in 4.0. If existing materials are purchased as off-the-shelf items in 16.0 after a preliminary study of the tryout evidence, then the flow would be $3.0 \to 16.0 \to 5.0$ rather than $3.0 \to 16.0 \to 4.0 \to 5.0$. More of this later.

One very promising technique is computer-assisted instruction (CAI). It embodies all five characteristics of truly programmed instruction above and does not require the mediation of any human other than the learner (5). It is a subset of machine-instruction (16.0) and is slated to become so significant in the next decade that the author has elected to extrude it from 16.0 and give it independent recognition. In planning the program (3.0), a decision is reached to utilize CAI. Unless an existing CAI processor (language, compiler, documentation) is to be acquired or accessed, it will be necessary to develop such a processor $(3.0 \to 20.0)$ and insert it into a time-sharing computer system $(20.0 \to 23.0)$. The documentation, customarily in the form of manuals or through a training program provided by the computer organization, is used as the basis for developing the CAI course in the appropriate language $(20.0 \to 22.0)$. Aside from creating a CAI processor, a task reserved for only top-line computer system analysts and programmers, the major work effort will be in 22.0. In 4.0, inputs from 2.0 through 3.0 cause a sequence of events to occur as shown in Figure 2-9 which reach the lesson plan level (4.3) then halt if human-instruction is the method. This sequence of events continues past the lesson plan level, down through the teaching point level 4.3.1 and halts at the step level of detail 4.3.3 if programmed instruction is the method. However, if CAI is the method, the

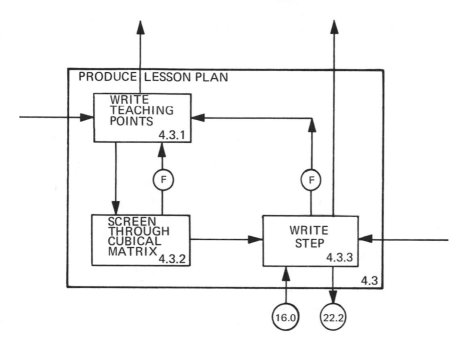

Figure 2-9. Third level of detail in the produce lesson plan *(4.3) subsystem.*

same sequence of events does *not* halt at 4.3.3, but the signal path outputs 4.0 and information flows to 22.0 where it is processed through CAI step development, the preparation of coding sheets to 23.0 and the production of the instructional program as CAI software. The production of software in 23.0 is the result of an input from 22.0 *and* another from 20.0 which by combination produces this new software in 23.0. A low-level simulation occurs in 23.0 as the instructional program is tried out, debugged and corrected by the Instructional Programmer taking the role of learner. Errors of various kinds are corrected as a result of the feedback loop 23.0 → 22.0 which controls two outputs of 22.0. Path 22.0 → 23.0 → 22.0 → 23.0 → 4.0 describes a program which passes the low-level simulation in 23.0 and is ready for a more formal tryout in 4.0. Path 22.0 → 16.0 → 4.0 describes how the specific CAI processor and step development in 22.0 determines various step formats of *innovative design* and input these to 16.0 where they are retransmitted to 4.3.3, shown in detail in Figure 2-9.

After a CAI program has been operationalized and a number of trainees have taken it, the training specialist will begin to develop a technical evaluation of the CAI processor's ability (20.0) to provide the behavioral outcomes required. These evaluation facts reach him through such feedback loops as $5.0 \rightarrow 4.0 \rightarrow 3.0$ and $6.0 \rightarrow 4.0 \rightarrow 3.0$. Since he is sitting virtually inside 3.0 making immediate decisions which are to affect future courses being planned, he is in a position to collect all factors which reveal where changes could be introduced into the CAI processor (20.0) and *expand* the language capability so that it can provide the behavioral outcomes he seeks. These alterations usually must be accepted by computer system management and can be instituted by following path $3.0 \rightarrow 20.0 \rightarrow 23.0$. However, if the computer software is under control of a separate management, and may even be in a different company, the alternative is to follow path $3.0 \rightarrow 21.0 \rightarrow 20.0 \rightarrow 23.0$. The request is fed back from 3.0 to 21.0 . . . a user's group. A *user's group* is a formal organization made up of users of various computing systems to give the users an opportunity to share knowledge they have gained in using a digital computing system and to exchange programs they have developed (5). The group also attempts to advise the software producer on changes they recommend in 20.0. The purpose of the feedback loop $3.0 \rightarrow 21.0 \rightarrow 20.0$ is to control two outputs of 20.0. Path $21.0 \rightarrow 20.0 \rightarrow 22.0$ controls the language specifications which form the basis of the manuals documented in 20.0, and input to 22.0. Path $21.0 \rightarrow 20.0 \rightarrow 23.0$ controls the processor software stored inside the computer.

Involved in and one basis for decision-making in 3.0 to utilize human-instruction, machine-instruction, CAI or some appropriate combination, is the formulation of learning psychology (14.0). There are many different theories in psychology which deal with human learning. All of these are represented in 14.0, where a particular alternative may be selected and applied in this manner:

1. $14.0 \rightarrow 15.0 \rightarrow 4.0 \rightarrow 5.0$: particular theory applied to human-instruction, tried out through simulation, validated and operationalized in a full-scale project.
2. $14.0 \rightarrow 15.0 \rightarrow 5.0$: particular theory applied to human-instruction and operationalized in a full-scale project.
3. $14.0 \rightarrow 16.0 \rightarrow 5.0$: particular theory applied to machine-instruction purchased from an outside source and operationalized in a full-scale project.
4. $14.0 \rightarrow 16.0 \rightarrow 4.0 \rightarrow 5.0$: particular theory applied to machine-instruction, tried out through simulation, validated and operationalized in a full-scale project.
5. $14.0 \rightarrow 16.0 \rightarrow 4.0 \rightarrow 22.0 \rightarrow 23.0 \rightarrow 4.0 \rightarrow 5.0$: particular theory applied to computer-assisted instruction, step format selected, steps prepared, translated into CAI steps, tried out at several levels of simulation, validated and operationalized in a full-scale project.

Beyond the microlevel aspect of the application of learning psychology, typified by the learner-teacher and learner-teaching machine relationships, is the macrolevel aspect dealing with psychology applied to facilities and equipment design and logistical support. Human performance engineering (17.0) is influenced by the particular psychological theory selected (14.0). It attempts to combine theoretical data with engineering state of the art to create learning environments which are comfortable, efficient, maintainable and durable so that learners can be processed through them expeditiously and at low cost without sacrificing the quality of instruction by targeting behavioral goals. The flow $14.0 \rightarrow 17.0 \rightarrow 19.0 \rightarrow 4.0$ delineates the design of proper facilities which are used in curriculum development. Inferentially, the same facilities are used in 5.0 to conduct the program. The flow $14.0 \rightarrow 17.0 \rightarrow 18.0 \rightarrow 19.0 \rightarrow 4.0$ describes the provisioning of logistical support for the training facilities. For example, if the training was technical and dealt with computer programming, the facilities (19.0) would include a computer, while logistical support (18.0) would be in the form of perforated tape, magnetic tape, punch cards, tab paper, equipment maintenance, etc., without which the training facility could not be operated. Anything needed to support the *facility*, including installed, fixed equipment, follows that flow.

Another flow 14.0 → 17.0 → 18.0 → 4.0 represents the influence of learning psychology on performance engineering as it affects the provisioning of curriculum materials which are not developed: ancillary training materials such as maps and cutaway models; consumable supplies used by trainees and staff; workbooks, text-workbooks and programmed instructional materials having a one-time use; etc. In this path, all decisions are made concerning the acquisition, utilization and maintenance of audio-visual aids and devices not categorized as installed, fixed equipment.

In this chapter, a flowchart model was created consisting of 23 subsystems. Each is linked by signal paths to at least one other subsystem. The most complex subsystem, *develop curriculum* (4.0), is linked to 10 other subsystems. Of 12 inputs to and outputs from 4.0, four are feedback signal paths. The entire model has 12 feedback paths testifying to its cybernetic characteristics. Of the 12 feedback paths, three are from *outside* the producer organization.

The typical producer organization was modeled in Figure 2-6 to the *third* level of detail. This is in contrast with the "training system" model of Figure 2-8 shown only to the first level of detail. Figure 2-9 represents one subsystem of Figure 2-8 at the third level of detail, suggesting it should be possible to draw Figure 2-8 to at least that level for all 23 subsystems. In the following chapter, we will continue creating the training system model to the third and fourth levels of detail.

3- a model
to the third level of detail

"Talking Through" the Training System

Up to now, we have

1. *Identified* 23 subsystems and described each graphically in five
   words or less, accompanied by a point-numeric code
2. *Related* the subsystems to each other using signal paths which
   revealed the direction of information flow
3. *Differentiated*
   a. regular signal paths ⟶
   b. feedback signal paths ◀━(F)━
   c. feedforward signal paths ━(FF)▶
4. Performed a low-fidelity simulation by "talking through" the
   model, function by function, until all were explained and
   understood

The next step is to again talk through or narrate the model,
which has been expanded in depth from 23 to 164 subsystems. It
should be borne in mind that this is a *management planning and
process model.* It is not intended to provide detailed instruction of

a "how to" nature. For example, subsystem 23.1 is described as *punch cards*. To learn the various types of cards and the many ways they are punched and verified, the reader must go to another source. Subsystem 4.4.1 is *conduct tryout*: there will be a discussion of the tryout concept and several alternatives but the detail at the nuts-and-bolts level will not be explained here.

In working through the larger model, refer continuously to the supplementary figure inserted at the back of this book. In order to regain a more general perspective, glance at Figure 2-8 occasionally.

In speaking of alternatives, it is interesting to note that this model is designed for a large number of alternative solutions at many decision points along the sequence of events beginning with analysis of the job (2.0) and ending with consumption of products and services (8.0). There are several points of view which may be adopted, either individually, alternatively or collectively:

1. At one point, for example, the decision might be to *select methods* (3.3) and the alternatives to *provide human-instruction* (15.0), or *provide machine-instruction* (16.0) or engage in *develop CAI processor* (20.0) as a prerequisite to developing CAI courses (22.0). These would be considered mutually exclusive.
2. When the decision is reached in 3.3 it might be to use 65% human-instruction (15.0), 33% coordinated slide-tape presentations (16.0) and 2% computer-assisted instruction (20.0 → 22.0 → 23.0), a delicate *mix* of methods rather than absolutely one or another.
3. When a subsystem is entered, such as *write teaching points* (4.3.1), one may elect *not* to do this, pass through the function and continue on to 4.3.2. However, the risk involved and the penalty for bypassing one function must be mentally traded off to produce an optimum solution. To *optimize* means to obtain the best results for the effort expended in terms of system output. It should be understood that optimizing in one subsystem at the third level of detail, for example

4.3.1, may not produce an optimum result in the output—in 6.4.3.

4. A subsystem with several inputs and outputs has relationships with other subsystems. One may elect to ignore some of these; for example in 4.4.1:

    a. Input from 4.3 could not be ignored since it represents the course content.

    b. Ignore input from 18.2 only if the course has no hardware component.

    c. Ignore input from 4.2.1 only if there are no pre- or post-tests or other tests.

    d. Ignore input from 18.1.3.4 only if there are no software consumables.

    e. Ignore input from 23.12 only if there is no CAI instruction program.

    f. Ignore input from 4.0 only if there are no ancillary materials such as workbooks and text-workbooks.

5. Even at the third level of detail, a subsystem may call for a large number of decisions within it and these result in many alternative solutions not covered in the model. To illustrate, *review effectiveness* (16.3.2) may require that the materials satisfy 20 to 30 different criteria, depending on the particular application. The *resolution* or number of levels of detail in this model is limited to four. While it is possible to analyze 16.3.2, or any subsystem, to produce a greater resolution it is beyond the author's scope in this volume. Resolution is a measure of the fineness of detail; lenses with greater resolution similarly illuminate microstructure. The equation $C = f(e,i)$ describes complexity (C) as a function of the number of elements (e) and the number of interrelationships (i). Since this is an acceptable formulation, increasing the resolution will reveal the microstructure (1). *It will not increase complexity but simply expose it.*

Refer in subsequent discussions to the supplementary figure.

	Object	Action	Information
	Main Group	Main Action	Main Fact
Task	Group	Action	Fact
	Unit	Step	Point
	Subunit	Substep	Subpoint

Figure 3-1. Hierarchy of stages in analyzing tasks into actions and informa-tion.

Produce Job Analysis-Human Activity Analysis (2.0)

We begin with existing employee behavior—various skills, knowledge and attitudes essential for performing activities in the ultimate work environment. *Analyze elements* (2.1) is the proce-dure for breaking down the whole job or human activity into smaller parts. A job consists of parts called tasks, and these are further decomposed or broken down into actions (manipulative actions and procedures performed by humans), facts (information) or groups (physical objects). A hierarchy appears in Figure 3-1. All of these parts are termed *elements.* Elements when added together constitute the *Universe* (*U*). This concept is expressed by the equa-tion $U = O + A + I$ where:

All objects = O
All actions $= A$
All information = I

The Universe is limited to a particular subset, specifically the particular job under study. Therefore $O, A,$ and I refer only to the elements in *produce products & services* (6.1) shown by path 6.4.1 → 6.1 → 7.1 → 7.2 → 2.1. Most knowledge about an existing job is well organized and categorized by the time a decision is made to set up a training program. However, this knowledge may be diffi-cult to obtain unless the analyst has had personal experience in performing the job. Personal experience or expertise is the conventional resource. In the absence of this, the analyst must substitute the experience of others and follow path 6.4.1 → 6.1 → 7.1 → 7.2 → 2.1, or a similar one.

The experience of others may be obtained by:

1. Observing activity while it is performed (6.4.3 → 6.4.1 →. . . 2.1)
2. Interviewing employee who performs (6.4.3 → 6.4.1 →. . . 2.1)
3. Interviewing employee's first-line supervisor (6.4.1 →. . . 2.1)
4. Interviewing non-supervisor expert, such as company industrial engineer (6.4.2. → 6.4.1 →. . . 2.1) or university faculty member (10.0 → 6.4.2 → 6.4.1 →. . . 2.1)
5. Discussing with a colleague (6.1 →. . . 2.1)
6. Reading book or other literature produced by expert (10.0 → 6.4.2 → 6.4.1 →. . . 2.1)
7. Interviewing employees by questionnaire (6.4.3 → 6.4.1 →. . . 2.1)

Note that first-line supervision (6.4.1) is always embedded in the signal path flow, emphasizing the importance in American business and industry of *never* bypassing this group, particularly while analyzing job content.

Each element in 2.1 can be related to real-life. *Establish DIG relationship* (2.2) describes whether the element is *directly* related (*D*), *indirectly* related (*I*) or *generally* related (*G*). Elements which are *unrelated* (*U*) never appear in 2.1 since they would act as distractors and reduce the overall efficiency (*E*) of the analysis.

Silvern's updated equation to describe the mathematical relationship, first published in 1957 (25), is:

$$E = 10D + 5I + G - 2U,$$

suggesting that *D* elements are *essential* and the job cannot be performed if *D* elements are missing; *I* elements are *important* and the objective is accomplished better and with more complete understanding by the employee; *G* elements are background or

Value of Level	Meaning of Value
10	Must perform perfectly, all of the time; without difficulty, and without any assistance
9 8 7	Must perform excellently most of the time; may use performance aid or supervisory assistance occasionally
6 5 4	Must perform very well most of the time; may use performance aid or supervisory assistance moderately
3 2 1	Must perform well most of the time; may use performance aid or supervisory assistance frequently
0	Not to perform at any time

Figure 3-2. Level of proficiency scale.

nice to know and contribute in breadth but never in depth. The job can be performed quite well without *I* or *G* elements. It is degraded if *U* elements are present.

Thus 2.2 screens all of the elements in 2.1, assigning a *D,I* or *G* to each, using analyst judgments to establish those which are essential and those which are not.

Quantify level of proficiency (2.3) is an effort to specify how well an employee must be able to perform each element in 2.1. Figure 3-2 shows an arbitrary level of proficiency scale.

The values in Figure 3-2 are slightly ambiguous. After all, different supervisors will not agree on the meaning of "occasionally." It may mean once an hour or twice a week, depending on specific environments and circumstances. In fact, a glance at 6.4.2 will reveal that most supervisors have these standards in their minds and can easily change their minds: 6.4.2 → 6.4.1 → 6.4.2. The analyst must refine the scale in Figure 3-2 and apply it in 2.3 based on supervisory agreement in 6.4.1. The concept in 2.3 is to devise a practical scale so that each element in 2.1 can be identified. When the training program is developed, elements identified as LOP = 10 will be instructed repetitiously (4.3.1) and tested more precisely (4.2) than those with LOP = 2, for example. The term *criticality* has been applied to this concept; elements with high LOPs are more critical for job success than those with low LOPs.

The analyst must exercise care in conducting the analysis (2.0). It would be impossible to explain an element rated *D* having an LOP = O! Obviously, the resources used were in total disagreement, one believing the element to be essential, another convinced it was not to be performed at any time! More frequent are cases where the element is rated *I* with an LOP = 9. The element is important and the employee is expected to perform it excellently most of the time with only occasional assistance. The analyst may be expected to determine such additional parameters not shown in the model as:

1. Frequency. Number of repetitions on a time scale (seconds, minutes, hours, days)
2. Suddenness or onset. Time (seconds, minutes) from awareness of need to act until action time has expired
3. Completion. Time (seconds, minutes) from beginning a continuous action until completing it
4. Accident proneness. Employee and/or property hazardness aspect of an action
5. Team sharing. Concurrency and transfer of action to team member

In many situations, time is built into a parameter and it often enters significantly into the trade-off decision in 2.3. *Quality* is, of course, basic to performance but competes with *time* in a technological society which tends to convert time into dollars. Since time is money, we like to reduce the time required to perform a task. When time involves human life, there is even greater interest in measuring it in 2.3. The job of teller in a savings and loan association, involving error rate permitted when adding checks on a desk calculator, can be compared with the job of commercial airline pilot involving error rate allowed in lowering landing gear.

Generally, the analyst must rely on the standards (6.4.2) under which the supervisor (6.4.1) functions. Large companies have staffs of industrial engineers who establish various work standards. These, when combined with supervisory standards, are aiming points for the job analysts.

In the process of extracting these elements from the job environment (7.1), the analyst may question a procedure or information obtained from a person or document. It is expected that this question will be raised in 7.2 to prevent an erroneous element from reaching 2.1, where it might eventually follow the path $2.1 \rightarrow 3.1 \rightarrow 3.2 \rightarrow 3.3 \rightarrow 4.1.1 \rightarrow 4.1.2 \rightarrow 4.1.3 \rightarrow 4.3 \rightarrow 4.4.1 \rightarrow 4.4.2 \rightarrow 4.5.1 \rightarrow 4.5.2 \rightarrow 5.1 \rightarrow 6.1 \rightarrow 6.4.3$. At the last subsystem, it will be detected by the supervisor *after* a number of employees have been trained, and the effort will be to correct it as a malfunction in the system 6.4.3 rather than to screen it as questionable in 7.2, before it enters 2.1. There is always a danger that the analyst will assume responsibilities of the industrial engineer in 7.1 and modify the elements. As stated in Chapter 2, in theory the analyst should not be expected to create new job procedures or set job standards of employee performance. Questionable elements follow the path $7.2 \rightarrow 6.4.1$, where supervision decides the degree of error, if any. The feedback path controls supervisory performance in several ways:

1. $7.2 \rightarrow 6.4.1$ if the first-line supervisor is the resource person in error
2. $7.2 \rightarrow 6.4.1 \rightarrow 6.4.3$ if the employee is the resource person in error

3. 7.2 → 6.4.1 → 6.4.2 if the industrial engineer is the resource person in error
4. 7.2 → 6.4.1 → 6.1 if a colleague somewhere in the organization is the resource person in error

There is no practical manner for controlling errors output from 10.0, and the model does not include any quality control outside the organization. One final observation is germane; if a contractor provides services to an organization, it is common practice to station a customer engineer inside the organization. This is practiced by telephone companies, computer firms and other special-purpose technical firms. These outsiders, represented usually by service or customer engineers, are to be categorized as "colleagues" for the purpose of these discussions. If the organization being modeled is a military installation, civil service employees are categorized along with all levels of military personnel, *but* any non-civil servants are known as colleagues. This distinction is important so first-line supervision can retain both authority and responsibility at their operating levels.

Perform Basic Analysis (12.0)

There is a fundamental process of analysis which allows an analyst to breakdown or decompose anything into its smallest bits and pieces (26). Objects of any dimension can be analyzed into main groups, groups, units and subunits (see Figure 3-1). Actions can be broken down into main actions, actions, steps and substeps. Information can be analyzed into main facts, facts, points and subpoints. As jobs in the United States become more mental and less manipulative, the content of a job will shift from objects and actions to information as the basis of content. For example, the task of filling a tooth might contain 30% object, 30% action and 40% information content; the task of taking a "Buy" order and entering it on a CRT terminal in the branch office of a brokerage house may have 1% object, 4% action and 95% information content. This shift to information is subtle but it is a covert measure of a major movement to reduce human manipulative activity and also to rely on machine information processing as an assist.

Regardless of what is to be analyzed, or its level of complexness, it can be broken down through use of the model in subsystem 12.0. The process is:

1. Identify the element, initially the whole (12.1).
2. Identify the major elements constituting the whole (12.1).
3. Try to relate major elements to each other (12.2).
4. Because the whole exists, relationships between major elements are not difficult to deduce in 12.2.
5. Each time, also relate the elements to the greater whole (12.2)
6. Separate the elements (12.3).
7. Limit the separation (12.4) to avoid losing identity.
8. Decide (12.4) if the level of detail is adequate for the purpose in 2.1.
   a. If adequate, then 12.4 → 2.1
   b. If requires deeper level of detail, then 12.4 → 12.1 → 12.2 → 12.3 → 12.4 when adequateness of level is again to be decided.
9. Continue iterating, 12.4 → 12.1 → 12.2 → 12.3 → 12.4, until the analyst is satisfied with the adequateness of the level of detail, then 12.4 → 2.1.

Limiting is actually under the control of two factors. One is the *structure* of the object, action or information, which is often man-made and may be very complex or quite simple. The other factor is the analyst's *decision* to limit based on how detailed he wishes to make his analysis, which may also be a function of the time he has allocated for this effort. It should be obvious that 2.1 can be performed only if the events in 12.0 occur. The result is a document called the job analysis-human activity analysis or JA-HAA.

Plan Program (3.0)

When the JA-HAA (2.0) has been produced it is necessary to *establish operating parameters* (3.1). One set of parameters deals with selection criteria for the trainees. If the trainees are employees of the company, some of the parameters, in addition to personal need for growth and supervisory approval or nomination, are:

1. Title or occupational family grouping on a career ladder
2. Hierarchical level on a career ladder, such as GS-13 in the U.S. Civil Service
3. Paygrade or salary group
4. Longevity of service or seniority in company and/or in paygrade
5. Union or trade association approval or nomination
6. Geographical location of employment
7. Prior job experience measured in years or by examination of proficiency ratings
8. Immediacy of training compared with availability of trainee

In some cases, trainees may be employed by customers and enrolled in a customer training program. The training organization might have less selection control and be limited to establishing:

1. Prospective trainee need to acquire particular skill and knowledge for product support
2. Entry level requirements or course prerequisites
3. Lead-time to date trainee is expected to apply course content on the job
4. Availability of prospective trainee

In other instances, trainees may not be employed either by the training organization or by a customer. The relationship between trainer and trainee may be contractual. Colleges, trade schools, institutes, special-purpose courses and others simply establish unique criteria in 3.1.

Another set of operating parameters in 3.1 has to do with management-centered outcomes. What specifically does management expect of the program? Only critical items should be included since they become design criteria as they move from 3.1 to 3.2. Here are just a few typical parameters in this set:

1. Providing a work force able to furnish certain amounts of manpower in specific occupational areas, measured in man-hours, man-days, man-months or man-years
2. Maintaining a work force in equilibrium able to furnish candidates in a particular occupation qualified to be considered for promotion to the next superior level on a time scale measured in weeks, months or years
3. Providing a work force able to furnish candidates for lateral transfer to related or semi-related occupations on a time scale measured in months or years
4. Providing a marketing organization able to communicate with the design organization and convey in specific, unambiguous technical language precisely what the market desires and is willing to purchase
5. Providing third echelon management of a multidivisional company with corporate visibility so that communications between operating divisions will be improved in the area of product design

Yet another set of operating parameters in 3.1 falls midway between trainee selection criteria and management goals of a semi-global nature. These are more mundane and refer to such matters as:

1. Employee training cost allocations in terms of pay during training, transit costs per diem, tuition or training costs chargeable to general overhead or special account
2. Temporary employee replacement pools to maintain production levels during transit and training period, including break-in time
3. Permanent employee replacement to maintain production levels when program is designed to promote or transfer in the organization
4. Employee assignment during slack work periods to reduce disequilibrium

Large companies profit through this type of thinking in 3.1, but even very small firms can be managed more effectively using the same model by simply establishing realistic parameters which work for them. There is nothing more useless than someone else's parameters!

Once the parameters have been established in 3.1 they are quantified in 3.2. Quantification is the process of assigning numerical values and units of measurement to a parameter. Let us say in 3.1 that the parameter is "immediacy of training compared with availability of trainee." In 3.2, quantification could result in a statement such as, "Trainee must be available and assigned within five days after distribution of announcement; course begins 10 days after announcement; trainees unable to be assigned but otherwise qualified are held in a queue for next course offering."

The danger in *not* quantifying is always greater than assigning numerical values and units. This danger increases with the size and complexity of the organization. Another danger is decision-reversal by senior management to favor one section or department. If quantification is well done, there should be no extenuating circumstances which might cast a shadow over the administrative competence of the training organization.

Here is another illustration of quantifying a parameter in 3.2. The parameter is "lead-time to date trainee is expected to apply course content on the job in his organization." Quantification results in the announcement that "Trainee is expected to apply course content on the job within 20 days; trainee's level of proficiency will drop 50% in 21 to 100 days of disuse." Of course, not all tasks suffer radically from a delayed application of instruction, but there will always be a need to quantify any delay in transfer from course to real-life.

Selection parameters and their values in 3.2 move to 13.1 and are refined and formally established as operating policy. *All* parameters and values input (3.3) where the first *rough cut* decision is made to select one or more methods of instruction. Suppose in 3.3 that the decision is to provide human-instruction (15.0). It is not necessary to specify 15.0 as a straight lecture, conference group, workshop, colloquium, symposium, laboratory or any combination of these. A specific decision at that level of detail follows the path $3.3 \rightarrow 4.1.1 \rightarrow 4.1.2 \rightarrow 4.1.3 \rightarrow 4.3.1$, where an input from 15.0 to 4.3.1 calls for a *final cut* decision. By that time, various mixes can be designed to optimize the instruction. Rough cut decisions which can be made in 3.3 are:

1. $3.3 \rightarrow 15.0$
2. $3.3 \rightarrow 16.1$
3. $3.3 \rightarrow 20.1$

These are predicated on principles and theories in the field of learning psychology (14.0) expressed as $14.0 \rightarrow 3.3$. For example, a training program being planned is expected to alter the attitudes of Caucasian prison officers with a rural, lower middle class back ground towards minority offenders with urban, ghetto cultural backgrounds. One objective is to have the officers understand the rules of behavior in a ghetto subculture. Research in 14.3 might indicate an encounter or sensitivity T-group method of instruction decision in 3.3 conducted in 15.0 by a black urban sociologist. This would be in sharp contract to an alternative solution of having the same sociologist lecture, and in even greater contrast to having the officers take a programmed course covering Black history.

When a decision is reached in 3.3, a signal path to 18.1.1 starts up the logistical support function so that software and hardware can be designed, produced and delivered following the path 3.3 → 18.0 → 4.4.1. Also, a signal path to 17.1 alerts the human performance engineering staff (17.0) so that it may analyze man-machine relationships in those programs incorporating complex instrumentation such as simulators, actual equipment, etc.

The most common form of instruction in the United States is *human-instruction* (15.0) and, as a subset, straight lecture on virtually a one-way basis. Not necessarily the best, it is nevertheless prevalent, often because it is least expensive and may not require any major preparation effort. The decision-making process in 3.3 must also consider the cost, usually expressed in dollars per trainee per hour, and preparation cost and time. Figure 3-3 depicts the operating cost per trainee hour including preparation but excludes salaries for trainees (5). These costs have risen quite sharply since they were first computed several years ago. In business and industry, unlike public schools or universities, the typical trainee is in a pay status while attending a training program. Therefore, it is to the financial advantage of the company to *reduce* trainee time even if it means initially investing more money in the preparation phase.

If the decision is to use programmed instruction which is not available off-the-shelf but is to be produced, the development cost (C_D) in dollars per hour of trainee exposure is expressed by:

$$C_D = \frac{C_{TP} + C_M}{t}$$

	$/Trainee/Hour
Elementary-Secondary	$0.16–$0.46
Publicly Controlled University	$0.81–$1.01
Publicly Controlled Professional School	$1.93
Military Training Programs	$0.36–$7.10

Figure 3-3. Range of costs per trainee per hour.

where: C_{TP} = Cost, total production labor

C_M = Cost, materials

t = time, hours of total trainee exposure

In 1970, C_D for one trainee hour of typical programmed instruction was $645 (5). Also in 1970, the cost, C_D, of producing one hour of computer-assisted instruction was $1450 using a commercial time-sharing system, validation based on a ten-trainee tryout, and a teleprinter console. C_D with a CRT console was $1667. The C_D for a teleprinter console with an audiovisual component (projected slides or prerecorded magnetic tape) was $1694. In other words, a very sophisticated version of CAI could be used for one hour of trainee instruction at a development cost of only 2.5 times that of a programmed instruction text. When these development costs are amortized among the trainee group, the actual cost per trainee drops dramatically.

Provide Human-Instruction (15.0)

Some attention has already been given to the Provide Human-Instruction subsystem. It is generally agreed in business, industry and government that technically based subject matter is best instructed by a specialist in his technical field rather than by someone who must first learn the subject matter. It has been said, "If you want plumbing taught, find a good plumber and teach him how to instruct—don't teach a good instructor to be a plumber—it'll never work."

Historically, the training specialist was first a subject matter specialist performing regular work in his organization; when a requirement arose to train employees, he was selected and temporarily transferred to conduct the program (27). In many instances, this temporary arrangement became permanent. The main task was instruction. In some firms, the arrangement continued as temporary and employees were *rotated* through the assignment. It was often considered punishment for rubbing management the wrong way! If not used as payment for infidelity, it became an elephant cemetary where aging beasts were abandoned to expire. Today, the training specialist in smaller companies is still looked to as an instructor, providing services in 15.0. In larger firms, he may enter with a communicative role and move to curriculum planning and development assignments. In any case, he relies on 14.0 to assist in writing lesson plans (15.0 → 4.3.1). Usually the lesson plan is written only to the teaching point level if used for human-instruction. An innovative director heading the training organization would expect the instructor (15.0) to write the plan (4.3.1), conduct a *tryout* (4.4.1), analyze the results (4.4.2), apply performance criteria (4.5.1) and accept or reject the plan (4.5.2). This is accomplished *before* the lesson plan and the instructor input (5.1), where *regular* instruction begins. Tryout of any instruction is a form of simulation, using samples of the target trainee population. It assures the organization that training is effective and trainees will transfer to their jobs in 6.4.3 what is predicted and expected. There are several possibilities:

1. Tryout is successful: 4.4.1 → 4.4.2 → 4.5.1 → 4.5.2 → 5.1 → 6.1.

2. Tryout is unsuccessful: 4.4.1 → 4.4.2 → 4.5.1 → 4.5.2 → feedback to 3.4 → 3.5 → 3.3. The feedback signal from 4.5.2 to 3.4 (*compare with parameters*) causes the results of the tryout to be compared with the initial sets of design parameters from 3.1 and 3.2. Notice that the model in the supplementary figure does *not* permit modifying or otherwise tampering with trainee selection criteria. The principle expressed in this particular model is that once the trainee pool is identified, instruction must be modified to meet trainee and company needs—and not the reverse.

When tryout is unsuccessful, either the *method* of instruction, or the *content* being instructed, or both, must be modified 3.4 → 3.5 → 3.3 → 15.0 → 4.3.1 → 4.4.1 → 4.4.2), again reexamined in 4.5.1 and accepted or rejected in 4.5.2. Feedback always controls the output of the function it inputs. This is a quality control loop which tries out a course and validates it *before* it is placed into widespread use. Even after it is operated in 5.1, *evaluate course* (5.2) has a feedback loop 5.2 → 5.1 to continue improving the course until it is finally eliminated. Those who complete the course follow the path 5.1 → 6.1 → 6.4.3 → 6.4.1, where the first-line supervisor perceives and evaluates job performance using standards from 6.4.2. Assume a supervisor is dissatisfied with a training program graduate's performance—it doesn't meet his standards. He feeds back an error signal 6.4.1 → 4.6 to the training organization and a sequence of events occurs to correct the error. A number of possibilities exist:

1. Supervisor (6.4.1) might have an unreasonable standard (6.4.2) unable to be satisfied by the training program; this calls for management control or supervisory self-control (6.4.1 → 6.4.2).
2. Supervisor (6.4.1) might have detected incorrect information or procedures calling for reexamination of course content and possibly the original job analysis using feedback loops: 6.4.1 → 4.6 → 4.1.1 → 4.1.2 → 4.3.1 → 4.3.2 → 4.3.3 → 2.1, or 6.4.1 → 4.6 → 2.1.

3. Supervisor (6.4.1) might have detected an employee personali-
    ty defect reducing his value in the job situation; this calls for
    supervisory correction (6.4.1 → 6.4.3) as feedback to control
    the output of 6.4.3. If groups of employees have such defects
    after training, 3.1, 3.2 and 13.1 need to be adjusted, but the
    model in the supplementary figure is not designed for that
    feedback contingency.

Provide Machine-Instruction (16.0)

It is possible in *select method* (3.3) to decide that a machine
will be used as instructor. This could be a cassette recorder with
prerecorded magnetic tape, closed-circuit TV with video tape,
teaching machine with two-way interactive software, programmed
instruction text with sliding mask, film projector using 16mm film
or any device which may be used by one trainee or a group. This
decision in 3.3 does not always reflect a philosophical *choice*
between man and machine but might ensue from this reasoning:
1. Trainees are remotely located in distant offices; cost of trans-
    porting to central location or sending instructor would be
    excessive.
2. Trainees cannot be spared due to work peaks for more than 30
    minutes at a time but can find quiet area near work station for
    two or three 30-minute sessions each day.
3. Trainees are in decentralized emergency service and stand by
    equipment awaiting alarm; while in standby they have ample
    time for machine-instruction.
4. Trainees are all from one department and instructing them
    simultaneously would shut down the entire department, yet
    instructing in very small groups would raise training cost.

It usually winds down to a trade between the cost of training
and the effectiveness of training. Management will tend to accept a
decision providing for maximum effectiveness at lowest cost.
Regrettably, the facts supporting one decision or another are
fuzzy at times and the trainees suffer from a poor alternative
solution.

Generally speaking, machine-instruction does not require the mediation of an instructor and thus differs from human-instruction. However, materials are now being produced for use solely by individual trainees, by groups of trainees without an instructor and by instructors as an aid to their conventional instruction. These are differentiated as:

1. *Teaching machines* or programmed instruction devices which incorporate
   a. Participative, overt interaction or two-way communication between trainee and machine or program
   b. Sequences which are carefully controlled and consistent
   c. Immediate knowledge of progress by trainee as feedback signals to control his subsequent behavior and reinforce it
   d. Pacing under trainee control
   e. Learning as the objective (acquisition, retention, transfer)
2. *Performance aid*
   a. Overt, one-way communication from machine or program to employee
   b. Sequences which are carefully controlled and consistent
   c. Pacing under employee control
   d. Performance as the objective (acquisition, transfer)
3. *Communications aid*
   a. Covert, one-way communication from machine or program to trainee
   b. Sequences which are carefully controlled and consistent
   c. Pacing under program or machine control
   d. Learning as the objective (acquisition, retention, transfer)

The performance aid is utilized in factories and offices where a long sequence of relatively complex events must be performed by an employee without omission. A tape recorder or visual projection display is used in a step-by-step program (23). It avoids reliance on human memory by furnishing a machine substitute for memory. The communications aid is well known as the audio-

visual aid renamed to eliminate any confusion with the other categories which are also audio-visual. These three categories should
cover all types of machine-instruction in 16.0.

It is assumed for purposes of illustration that the decision in
3.3 is to use a programmed text exclusively and that there will be
no other instruction of any kind. Using the JA-HAA documentation output from 2.1, the path 2.1 → 3.1 → 3.2 → 3.3 → 16.1 is
followed. A decision must be made in 16.1 to make or buy the
programmed text. Here are some of the alternative paths:

1. *Buy:* 3.3 → 16.1 → 16.2 → 16.3.1 → 16.3.2 → 5.1. A purchase
    will be made to procure off-the-shelf (16.2). There is a danger
    in buying programmed materials simply by reading a one-para-
    graph description in a flyer or a directory. These materials in
    theory, and hopefully in practice, have been tried out and
    validated. Unlike an ordinary text, they are supposed to
    produce genuine behavioral changes (28). Consequently 16.2
    outputs to 16.3.1, where these behavioral objectives are re-
    viewed in comparison with the behaviors specified in the
    JA-HAA by inputting 2.1 to 16.3.1. While 16.3.1 does not
    graphically reveal a reject or abort function, it clearly exists
    since much of what is available on the commercial market is
    inappropriate to the specific needs of a training organization
    when compared with the elements of a carefully prepared job
    analysis. In 16.3.1 the comparison deals with the *content* of
    the programmed text. In 16.3.2 it compares the *effectiveness*
    of the text. This is measured by pre-test/post-test gains to be
    certain that time spent by the trainee produces learning which
    can be acquired, retained and transferred efficiently to the
    real-life (job) environment. Here again, a reject or abort func-
    tion is present since much commercially available material is
    ineffective for adults in employee training programs. If the test
    satisfied the reviews in 16.3 it moves directly from 16.3.2 to
    instruct course (5.1). Having already been tried out and vali-
    dated by the commercial publisher, it need not trace a path

through 4.4 and 4.5 to 5.1. The training specialist must always ask the publisher for tryout and validation data. He should also anticipate a response suggesting that there isn't any.

2. *Make by contracting out-house:* 3.3 → 16.1 → 16.4.1 → 16.3.1 → 16.3.2 → 5.1. To contract "out-house" means to issue a contract to an outside firm for producing, trying out, validating, debugging and delivering a ready-to-go programmed text which meets specifications of the contracting agency. This is contrasted with "in-house," wherein all functions are conducted inside the organization's training department. Squeamish individuals have referred to 16.4.1 as "out-of-house" but the author continues to prefer out-house. In some instances, the training organization performs the job analysis (2.0) since it is less disruptive and submits it to the contractor for developing the text. This provides greater technical control by the company. However, the contractor may not agree to guarantee results unless he also produces the job analysis. The major problem is locating a contractor capable of creating reliable programmed instructional material at a reasonable price. By reasonable price is meant $645 plus costs and profit for a one-hour program. Using

$$C_D = C_{TP} + 0.5\ C_{TP} + 0.1\ C_{TP}$$

where C_D = Cost, development
C_{TP} = Cost, total production labor,

out-house contractor overhead equal to 50% and profit of 10%, the company should expect to buy a fully tried out and validated text for $1032 per hour of trainee-text interaction. However, if the text is to be used for 10 hours of instruction, the contractor may be expected to estimate $9000 and not $10,320 (5). Regardless of the length, the training specialist will review the content (16.3.1) and effectiveness (16.3.2), relying on an input from 2.3 before accepting and using it for instruction (5.1).

Activity	Man-hours
Development of course materials	2800
Art, tryout editions	90
Art, first editions	320
Typing, tryout editions	225
Typing, first edition	200
Total	3645

Figure 3-4. Production man-hours for a programmed course of 41 hours trainee time.

3. *Make by producing in-house:* 3.3 → 16.1 → 16.4.2 → 16.5 → 16.6 → 4.3.3 → 4.8 → 4.4.1 → 4.4.2 → 4.5.1 → 4.5.2 → 5.1. In-house activity consists of training organization design, development and execution of the programmed text project. Internal costs will vary as different salary schedules are involved, but here are some man-hour estimates which may be useful in guiding decisions (5):
   a. A programmed text required 3645 man-hours to produce, distributed as shown in Figure 3-4.
   b. *Effectiveness,* measured with pre-test/post-test techniques and resultant gains, was 0.89 (mean); a typical trainee would learn 89.0% of what was intended.
   c. *Trainee time,* measured by having employees keep a record, was 41.0 hours (mean).

Activity	Hours
Production, steps/hour (4.3.3) Production, minutes/step	8.0 7.5
Trainee rate/hour during tryout (4.4.1)	60.0
Production hours to produce one trainee hour ready for tryout (4.4.1)	7.5

Figure 3-5. Production rates for a programmed course.

d. One hour of instruction required about 89 preparation hours, including clerical effort.

It is important to know various rates of text production and trainee consumption when functioning in-house. Figure 3-5 shows that subsystem 4.3.3 activity is performed at the rate of eight steps/hour, assuming all previous inputs are present. This averages 7.5 minutes/step. A trainee can work through about 60 such steps each hour during validation in 4.4.1. Therefore, the training specialist who is writing steps or performing instructional programming in 4.3.3 works for 7.5 hours to produce one hour of tryout material.

Now, let us examine chronological production time for an entire sequence of events to produce one trainee hour (refer to Figure 3-6):

1. Sequences 2.2 → 2.1, 2.3 → 2.1, 4.1.1 → 4.1.2 → 4.1.3 and 4.2.1 → 4.2.2 require 16.0 hours.
2. Sequence 4.3.1 → 4.3.2 → 4.3.3 requires 7.5 hours.
3. Sequences 4.4.1 → 4.4.2 → 4.5.1 → 4.5.2 and 4.5.2 → 3.4 → 3.5 → 3.3 → 4.1.1 →. . . 4.3.3 require 16.5 hours.

Activity	Hours
Production of steps for one trainee hour, ready for tryout	7.5
Production of JA-HAA + COTP + TEST for one trainee hour, ready for tryout	16.0
Tryout and validation of materials for one trainee hour	16.5
Total	40.0

Figure 3-6. Production hours to produce one trainee hour of programmed course.

Therefore, it is seen that 40.0 hours, or about one work week, are necessary to produce one hour of validated, reliable programmed text. These are chronological hours, not man-hours.

When the in-house decision occurs in 16.4.2, the training organization next faces the *design materials* (16.5) network of alternatives. The example being used is predicated on the in-house production of a programmed text. Thus, *design programmed text* (16.5.1) deals with such matters as:

1. Recognition or recall
2. Expendable or reusable text
3. Black ink or colors
4. Extent and type of graphics (line drawings, halftones)
5. Number of copies, tryout and first editions
6. Printing production turn around time
7. Binding and mask design

However, these are rough cut decisions in 16.5.1 which require refining sequences in 18.0. Decisions in 16.5.1 make it possible to *establish step format* 16.6. A step is usually a rectangular area enclosed with a border and containing three kinds of elements: stimulus material presented to the trainee, response area for his recognition or recall and response, feedback material presented to him which provides the most correct answer and thereby reinforces his correct behavior or offers remediation. Notice that *formulate learning psychology* (14.0) inputs 16.6 and is one basis for unique step formats. Some decisions in 16.6 deal with:

1. Particular design of trainee response area and degree of involvement
2. Types of tools for trainee response (pencil, ruler, protractor, lettering pen, etc.)
3. Arrangement of steps (page segment sequential, page segment vertical, branching, etc.)
4. Ancillary materials (workbook, vellums, etc.)
5. Concealment device and degree of control to prevent review of immediately preceding steps.

These are final decisions and enter 4.3.3 where they are binding upon the step-writing functionaries. The sequence continues, then, $4.3.3 \rightarrow 4.8 \rightarrow 4.4.1 \rightarrow 4.4.2 \rightarrow 4.5.1 \rightarrow 4.5.2 \rightarrow 5.1$.

Decisions Involving Other Kinds of Machines in 16.0

The model in the supplementary figure allows for decisions other than merely the selection of a programmed text in 16.5.1. Even within 16.5.1 it is possible to design a multimedia course which transcends a simple text-like design. The following example describes such a course (26):

1. Programmed text-workbook with page segment vertical design controls and directs trainee.
2. Trainee is directed to respond by constructing his response with pencil in his text-workbook.

3. Trainee is directed to turn crank of hand-held 8mm battery-powered viewer, view a single frame, view a series of motion sequences in color, or view a series in reverse and single out a specific frame, respond by writing into his text-workbook, receive feedback by sliding the mask.
4. Trainee is directed to place ¼-inch prerecorded magnetic audio tape onto a tape device and listen to an audio narration, respond by writing in his text-workbook and receive feedback by sliding the mask.
5. Trainee is directed to use a screwdriver and disassemble or assemble a complex mechanical part, respond by writing and sketching in his text-workbook and receive feedback by sliding the mask.

It should be clear in examining the model in the supplementary figure that the decision in 3.3 to select a multimedia programmed course as described above also triggers various sequences in *operate logistical support* (18.0) which result in outputs to 4.4.1 and inferentially to 5.1.

The characteristics of a teaching machine have been outlined earlier. One feature is concealment which prevents the trainee from looking ahead or behind and thereby controls his activity without prohibiting response. Depending on the particular subject matter, these are several considerations:

1. General-purpose teaching machine which can accommodate a wide variety of programs including the one being produced in-house (4.3.3)
2. Special-purpose teaching machine designed to instruct specific tasks only; may simulate the work environment to a high degree of fidelity
3. Storage and access capability (film, strip, slides, microfiche, aperture cards, audio tape, video tape, etc.)
4. Scoring capability for tryout application
5. Display area and method (visual, audio, audiovisual)
6. Synchronization of audio with visual (reliability)

7. Maintenance (spare machines, spare parts and service, reliability of components, time to repair)

Because the model has described implementation of human-
instruction (15.0) it is appropriate to explain decisions which
involve machines and software (16.5.3) which are used mostly to
supplement instruction. The instructor or, in large programs, the
training specialist must decide if any hardware or software exist
which will clarify his communication to the trainees. The list is
much too long to be included here. The decision is made in 15.0
and implemented in 4.3.1. At the complexity level where a film is
to be *produced,* and not simply procured off-the-shelf, it is made
in this sequence: $3.3 \rightarrow 16.1 \rightarrow 16.4.2 \rightarrow 16.5.3 \rightarrow 16.6 \rightarrow 4.3.3 \rightarrow 4.8 \rightarrow 4.4.1 \rightarrow 4.4.2 \rightarrow 4.5.1 \rightarrow 4.5.2 \rightarrow 5.1$.

Sedlik (29) has created a cybernetic model in LOGOS language
dealing just with film and TV production incorporating 48 subsystems which would fit inside of subsystem 16.5.3. Sedlik's major
subsystems are:

1. Conduct script research (1.0)
2. Prepare script outline (2.0)
3. Prepare treatment (3.0)
4. Prepare script (4.0)
5. Accomplish physical production (5.0)
6. Prepare rough cut/interlock (6.0)
7. Complete final production (7.0)

It is quickly seen that inserting the Sedlik model inside of
16.5.3 would immediately result in six levels of detail. It illustrates
that filmmaking and TV production are complex communications,
and production is best handled by skilled filmmakers and TV pro-
ducers. Despite advertising claims to the contrary these activities
cannot be performed with great success by amateurs posing as
training specialists.

Decisions Involving the Use of Computer-Assisted Instruction

A *processor* is a computer program (software) used in compiling a source program that will produce, when completed, an execution of the objective function of the program or process; this generic term includes assembly, compiling and generation (30). A CAI processor is designed to accept programs written by the instructional programmer (4.3.3) and to convert these into trainee's programs (23.8). The executed program (23.12) is educationally the equivalent of human-instruction (15.0) or machine-instruction (16.0) when it has completed tryout (4.4) and validation (4.5) as shown in the supplementary figure.

If a CAI course is to be written using a *non-CAI language* such as BASIC,FORTRAN or PL/1,there is no need to develop the CAI processor in 20.0. The computer manufacturer furnishes an appropriate language and all programming is by individuals proficient in that language. The path is 3.3 → 2.2.1. While eliminating all of 20.0, it requires professional *computer* programming in 22.2 and 22.3. For this reason, it is wise to develop a course in a CAI language which calls for implementation of 20.0 but permits *instructional* programming in 22.2 and 22.3.

Most training organizations are willing to accept a commercially available CAI language such as COPI, COURSEWRITER II and III, LYRIC and PLANIT. The signal path would be 3.3 → 22.1 since all functions in 20.0 would have been performed by the producer or vendor of the CAI language.

A large company may wish to develop its own CAI language or make major modifications of an existing language, since it may have a computer facility staff able to do this. The first step would be to develop language specifications (20.1), which can be quite extensive and extremely technical. The training specialist must have experience in instructional programming in order to make realistic and practical recommendations. Cost is expressed in computer run time, storage, on-line connection, etc., and it is ludicrous to have the language do things which would be rarely used in the company training programs and also are expensive to program and maintain.

The processor may be a compiler, pre-compiler, interpreter or other software package. A *compiler* is a program-making routine which produces a specific program for a particular problem by determining the intended meaning of an element of information expressed in pseudocode, selecting or generating the required subroutine, transforming the subroutine into specific coding for the specific problem, assigning specific storage registers, etc., and entering it as an element of the problem program, maintaining a record of the subroutines used and their position in the problem program and continuing to the next element of information in pseudocode (30). *Write compiler* (20.2) is the function of developing software which will take an instructional program written in natural language on coding sheets (22.3) and transform this into a trainee's program (23.8) *without* the programmer knowing anything about computing or what is happening in the computer.

It is necessary to *produce compiler* (20.3) which means, in essence, that the compiler program is run iteratively in the computer and under various practical conditions to be certain it is reliable and effective. *Documentation* is the process of collecting, organizing, storing, citing and dispensing documents or information recorded in the documents (30). It includes problem statement, flowcharts, coding, operating instructions, record of changes, etc. The justification for documentation is not simply to have archives but to use them to make accurate and low-cost modifications one to five years after the compiler has been installed and operating. Job mobility in the computing field is unusually high, and programmers often leave behind a perfectly functioning processor which cannot be modified because their successors are unable to decipher the program. One need not emphasize job security and salary increases which result from non-documentation and similar subtle techniques of witholding critical information. The 20.4 documentation function is not a responsibility of the instructional programmer. He does have documentation requirements which are expressed in 2.1, 4.1.3, 4.3.3, 23.14 and so forth.

One end-product of 20.4 is a manual, usually prepared by the compiler-processor software writing group, which may be impossible for the average training specialist to understand. Nevertheless, it is the basis for him to *learn specific CAI system* (22.1). The best manuals are those written by training specialists who are also computer specialists, so long as their training expertise is not limited to computing, mathematics or engineering. The software group may also offer a formal training program (20.4 → 22.1) which goes beyond the manual and provides some experience in using the hardware/software computing system.

Thus, the compiler is produced and installed in the computing system (20.3 → 23.8). There is technical information available so that the instructional programmer can use the CAI language following path 20.4 → 22.1 → 22.2 →. . . 23.8.

With a computing system and a *CAI processor* (20.0), it should be possible to *develop CAI course* (22.0) and *produce computer program* (23.0). No two CAI languages are alike and this means every CAI system is different. The hardware is different, the operating procedures are not identical, the programming rules are dissimilar, etc. It is not like learning to drive a Dodge (Chrysler) and then immediately driving a Buick (General Motors) or a Mercury (Ford). Except for manual or automatic shifting, all cars are alike and the driver of any one can, with minor discomfort, drive any other. The CAI programs written for the IBM 1410 in COURSEWRITER I will not execute on the IBM 1500 CAI System or on any of the IBM 360 or 370 models using COURSE-WRITER III. These require translation, a manual operation tediously performed at a desk by a human using a pencil, eraser and coding sheets.

There is, however, one exception to this industry-wide practice of non-standardization. This is LYRIC, which was co-invented by the author (5). An important LYRIC design goal was that it be *machine-independent* and installable in computers produced by many different manufacturers. This was solved by reliance on FORTRAN, a relatively standard scientific language available on most large time-sharing systems. A LYRIC pre-compiler *automatically* converts the LYRIC language (source) to FORTRAN programs which utilize a set of LYRIC subroutines. FORTRAN programs are then executed within the constraints of each individual computer system. In some instances, the programs are stored as FORTRAN programs and are recompiled at run time. In other systems, they are stored in binary and require no compilation but run immediately upon command. The main differences are cost of recompilation and waiting time—usually measured in seconds or perhaps a few minutes. However, the important LYRIC feature is that a CAI program written for one computer system will execute on another with minor modifications; thus it is machine-independent.

In 22.1 the instructional programmer must learn *at least* two things:

1. Operation codes of the specific CAI language which he will use
2. Strategies which he is permitted to consider in connection with the operation codes or opcodes

Opcode is defined by Sippl as a command usually given in machine language (30). It is shorthand for *operation code*, itself defined as symbols that designate a basic computer operation to be performed. Sippl also used the term *mnemonic operation code* to mean writing operation codes in a symbolic notation which is easier to remember than the actual operation codes of the machine. All of these terms originated in the early days of computing (the 1950s) when programming was performed in a machine-language or code understood by the computer. As computer software grew in sophistication, the computer programmer used mnemonic opcodes which looked more like natural language (English) words. The software translated these codes into machine-language or equivalent. For CAI, opcodes should be derived from natural language since the instructional programmer is not expected to know computing or computer languages other than his CAI language.

Some CAI languages have single-word opcodes, but others have symbols or groups of words or expressions which are equivalent to

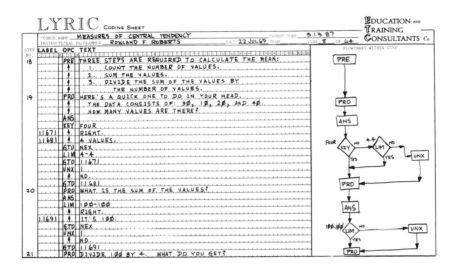

Figure 3-7. Typical opcodes (OPC) and strategy represented by flowchart within step. These appear on coding sheet (22.3).

one or more opcodes. Regardless, all of these are referred to here as opcodes. This is because an opcode in any form is a command which the computer will understand and accept. These commands constitute the step-by-step ground rules for writing CAI programs for the particular CAI system and must be learned. A step produced in 4.3.3 is converted into opcodes in 22.2 following an instructional strategy as in Figure 3-7. The path is 4.3.3 → 22.2 → 22.3. The strategies and opcodes dictate the step format 22.2 → 16.6 → 4.3.3 → 22.2 which closes the loop on 22.2.

The strategy is based on and constrained by the kinds of steps and the opcodes of the specific language. *Strategy* is defined here as the plan or scheme established in advance of the actual activity. In mathematical game theory, a pure strategy is one wherein the player chooses in advance the same alternative in each play of the game; the choice of different alternatives would constitute a mixed strategy. The strategy is represented as a flowchart model in Figure 3-7. It is probabalistic since the instructional programmer assumes certain characteristics of the trainee, then creates a plan to elicit a prescribed behavior. He tries to assign a high probability of successful outcome (win) by careful design. If he is wrong, the behavior is not elicited in a time/effectiveness frame of reference (lose). He first examines the outcome in 23.13 with himself as the learner and next in 4.4.1 with a sample of real students as learners taken from the target trainee population before the course is released for use in the company (5.1). If his strategy within a step (4.3.3) or between steps (within a teaching point [4.3.1]) is bad (lose), he follows paths 4.3.3 → 4.3.1 → 4.3.2 → 4.3.3 → 22.2 → 22.3 → 23.2 → 23.7 → 23.8 → 23.12 → 23.13. If 23.13 succeeds, he will document 23.14 and follow 23.12 → 4.4.1 → 4.4.2 → 4.5.1 → 4.5.2 → 5.1.

All of the foregoing explanation is presented to emphasize that the strategy is dependent upon specific opcodes and step kinds. Computing science state of the art does not yet permit opcodes to be generated as the result of a programming strategy except via path 3.3 → 20.1 → 20.2 → 20.3 → 23.8 and 4.3.3 → 22.2 → 22.3 → 23.2 or similar → 23.7 → 23.8.

Subsystem 23.0 consists essentially of producing and debugging the CAI program after it is written and before it enters tryout in 4.4.1. There are four practical methods of converting the coding sheet into machine-understandable input:

1. Punch cards (23.1). Produce 80-column Hollerith cards or 96-column IBM cards, read these remotely into a telephone line through a modem (23.7) which inputs the computer for processing (23.8). An alternative is to send the cards to the computer facility where they are read in (23.9) and the program is processed (23.8). The latter alternative deprecates the remoteness concept of CAI.
2. Input at keyboard (23.2). Manually type in program on-line into telephone line through a modem (23.2 → 23.7 → 23.8). While remote, runs costs up since system is on-line; excellent for debugging and correcting by using an editor language.

3. Punch perforated tape (23.3). Manually punch tape off-line then transmit it on-line through modem (23.3 → 23.5 → 23.7 → 23.8); this is considered superior in cost and effectiveness to 23.1.
4. Record magnetic tape (23.4). Similar to perforated tape in utility. The storage is magnetic cassette and the transmission speeds are higher (23.4 → 23.6 → 23.7 → 23.8); this is considered superior in cost and speed to 23.3.

In some computer systems, it is possible to utilize all four of these methods simultaneously. The author believes the most effective combination is 23.2 → 23.7 → 23.8, and 23.4 → 23.6 → 23.7 → 23.8.

When a program is processed, the instructional programmer or his editing assistant will *list instructional program* (23.10) as shown in Figure 3-8. Errors due to typing or transmission are checked and corrected in 23.11. Errors due to strategy, opcodes or labeling are fed back (23.11 → 22.3) and entered on the coding sheet thereby closing the loop on 22.3 and controlling the quality of output. This loop deals only with the original or unexecuted program and not with the version translated into FORTRAN, lower order machine language or binary code.

To produce a trainee version of the program, path 23.8 → 23.12 → 23.13 → 22.2 is followed. The instructional programmer debugs the interactive, conversational version as a trainee (23.13). A typical trainee program is seen in Figure 3-9.

Errors in 23.13 usually involve step construction and strategies within steps. These are controlled by feedback from 23.13 to 22.2. In effect, closing the loop on 22.2 controls the quality of output and refers to the trainee or executed program. When this program is ready for tryout, two events occur:

1. 23.13 → 23.14 (documentation)
2. 23.13 → 4.4.1 → 4.4.2 → 4.5.1 → 4.5.2 → 5.1 (tryout and validation)

```
695        GTO NEX
700        UNX 1
705        ' THE INSTRUCTION IS NOT OK.  IT IS CLEAR AND      OKAY.
710        ' COMPLETE, BUT HAS POOR SEQUENCE.
715        PRE THE THIRD REQUIREMENT OF THE PROCEDURAL STEP
720        ' IS THAT THE INSTRUCTION BE WRITTEN IN STRICT
725        ' CHRONOLOGICAL OR TIME SEQUENCE.
730        PRO NAME THE THREE RULES REGARDING THE CONTENT AND
735        ' STRUCTURE OF THE PROCEDURAL STEP?  MAKE YOUR ANSWER
740        ' AS SHORT AS POSSIBLE SO THAT IT WILL FIT IN ONE LINE.
745        ADD-C1,C1
750        ANS
755        KEY CL      CLAR
760        ADD+1,C1
765        GTO 46115
780 46115 KEY COMP
785        ADD+2,C1
790        GTO 46116
795 46116 KEY SEQ
800        ADD 4,C1
805        GTO 46117
810 46117 GTC C1,0/46118,46119,46121,46122,46123,46124,46125,46126
815 46118 ' SORRY, YOU MISSED ALL THREE. BUT......
820        GTO 46127
825 46119 ' YOU GOT CLARITY OK, BUT MISSED THE OTHER TWO.
830        GTO 46127
835 46121 ' COMPLETE IS ONE ANSWER.  YOU LACK TWO MORE.
840        GTO 46127
845 46122 ' CLEAR AND COMPLETE ARE RIGHT.  YOU NEED ONE MORE.
850        GTO 46127
855 46123 ' CORRECT SEQUENCE IS RIGHT, BUT YOU MISSED TWO.
860        GTO 46127
865 46124 ' CLARITY AND SEQUENCE ARE RIGHT; GET ONE MORE.
870        GTO 46127
875 46125 ' CORRECT ON CLDAR AND SEQUENCE.  YOU NEED ONE MORE.
880        GTO 46127
885 46126 ' EXCELLENT.  YOU HAVE ALL 3 ITUMS RIGHT.      RULES
890        GTO NEX
895 46127 UNX 1
900        ' KEEP TRYING.
905        GTO PRO
910        UNX 1
915        ' THE THREE ITEMS ARE CLEAR, COMPLETE , AND IN
920        ' STRICT CHRONOLOGICAL SEQUENCE.
925        PRE THIS IS THE END OF THE LESSON ON CONTENT AND STRUCTURE
930        ' OF THE PROCEDURAL STEP. NOW TAKE THE POST-TEST.
935        PRE
940        PRE               POST-TEST
942        PRE
944 46130 PRO 1. WHAT PSYCHOLOGICAL THEORY GOVERNS THE CONTENT
946        '      AND STRUCTURE OF THE PROCEDURAL STEP?
948        ADD-C2,C2
949        ANS
950        PAC . -
952        GUD/STIMULUS RESPONSE
954        ADD+5,C2
955        GTO NEX
956        PRO 2. WHAT ARE THE THREE RULES REGARDING CONTENT
958        '      AND STRUCTURE OF THE PROCEDURAL STEP?
960        ANS
```

Figure 3-8. Listing with errors circled.

(NEXT SLIDE --> 47)

WE HAVE ALREADY REFERRED TO THE INTERSECTION OF THE CROSS-
HAIRS AS THE CIRCLE. IT IS MORE ACCURATE TO CALL IT: IRIS

(NEXT SLIDE --> 48)

BEFORE PRESSING THE LOCK-ON SWITCH, THE PILOT MUST BE SURE
THE TARGET IS INSIDE THE -----.

? IRIS
YOU'VE GOT YOUR EYE ON THE BALL. (GET IT?)

THE TARGET APPEARS IN THE IRIS. THE PILOT PRESSES THE LOCK-
ON SWITCH. THE LASER IS MOUNTED INSIDE A LARGE GYROSCOPE.
AT THIS MOMENT, THE GYRO IS 'UNCAGED'....MEANING IT IS FREE
TO MOVE IN ALL FORWARD-LOOKING DIRECTIONS. THE LASER IS
LOOKING DIRECTLY AT THE TARGET. IF THE AIRCRAFT SHOULD
MOVE A LITTLE OFF TARGET, THE LASER WOULD BE LOCKED-ON THE
TARGET AND SWING AS THE AIRCRAFT TURNS AWAY.

SO, YOU CAN SAY THAT THE LASER WILL LOOK AT THE TARGET EVEN
IF THE AIRCRAFT IS LOOKING IN A SLIGHTLY DIFFERENT DIRECTION.
THE LASER, MEANING EYBOL, IS TRACKING THE TARGET.

FOR THE GYRO TO BE 'UNCAGED' AND TRACKING, WHAT ACTION BY
THE PILOT MUST HAVE OCCURRED AN INSTANT BEFORE?

? HE ACQUIRED THE TARGET VISUALLY
TRY AGAIN. FOR THE GYRO TO BE 'UNCAGED' AND TRACKING, WHAT ACTION BY
THE PILOT MUST HAVE OCCURRED AN INSTANT BEFORE?

? AN INSTANT BEFORE -- HE PRESSED THE LOCK-ON SWITCH
ROGER.

        (NEXT SLIDE --> 49)

WHILE TRACKING WITH EYBOL, THE PILOT IS ALLOWED SOME
VARIATION IN AIRCRAFT ATTITUDE. WITHIN THESE LIMITS, AND
FOR ANY COMBINATION OF THESE MOVEMENTS, THE LASER WILL
TRACK AND KEEP LOCK-ON. IF HE EXCEEDS THESE LIMITS, HE WILL
LOSE LOCK-ON. WHAT WILL HAPPEN IF HE YAWS 3 DEGREES?

? SINCE HE EXCEEDED THE LIMIT -- HE WILL LOSE HIS LOCKON
GOOD THINKING. HE LOSES LOCK-ON.

WHAT WOULD HAPPEN TO HIS LOCK-ON IF HE PITCHES 1 DEG, YAWS
1 DEG, AND ROLLS 1 DEG SIMULTANEOUSLY?

? STILL WITHIN TOLERANCE....NOTHING WILL HAPPEN
OKAY. SINCE HE IS WITHIN 1 DEGREE IN EACH ATTITUDE, HE WILL
REMAIN LOCKED-ON HIS TARGET.

SUPPOSE HE PITCHES 1 DEG, ROLLS 2 DEG, BUT DOES NOT YAW?

? EXCEEDS HIS ROLL LIMIT AND BREAKS LOCK-ON
BY ROLLING 2 DEG, HE LOSES LOCK-ON.

    (NEXT SLIDE --> 50)

WHICH COMBINATION OF VALUES (IN DEGREES) IS POSSIBLE DURING
LOCK-ON? (TYPE A, B, C, D, E OR F)

     PITCH   YAW   ROLL   FIELD OF VIEW
A. --- 1 --- 2 --- 1 ------ 2
B. --- 1 --- 1 --- 1 ------ 2
C. --- 1 --- 1 --- 1 ------ 3
D. --- 1 --- 1 --- 2 ------ 1
E. --- 1 --- 1 --- 1 ------ 1
F. NONE

? E
EXCELLENT SELECTION.

Figure 3-9. Typical trainee program.

User's Group Improve CAI Processor (21.0)

In working with a large CAI hardware/software system, a formal organization known as a *user's group* attempts to advise the software producer (20.3) on changes recommended in 20.1. The user's group is a formal organization made up of users of particular computing systems to give the users an opportunity to share knowledge they have gained from digital computing systems and exchange programs they have developed (30, 31). A danger in expanding a user's group to include many different computer systems is that time is devoted to general rather than highly specific recommendations in path 3.5 → 21.0 → 20.1. Obviously, a user's group which addressed itself to CAI in all computing systems is not a user's group but a society which has utility but lacks the leverage of a real user's group. The path 3.5 → 21.0 → 20.1 is to provide feedback through 21.0 and thereby control the output of 20.1.

In one sense, the user's group concept is much like the grocery store or supermarket cooperative concept with clout. In yet another, it is similar to the general model in Figure 2-6. The user's group is equivalent to *represent consumer attitude* (2.2), which feeds back recommendations to the software producer (2.1 → 2.2 → 1.3 → 1.1.1 → 1.1.2 → 1.1.3 → 2.1), closing the loop on 1.1.1 and controlling its output.

Operate Logistical Support (18.0)

We have traced signal paths following the logic for providing human-instruction (15.0), machine-instruction (16.0) and computer-assisted instruction (20.0, 21.0, 22.0 and 23.0). In all of these instruction methods, it should be painfully obvious that the instructor, training specialist, human resources developer—in fact the entire training organization—requires logistical support. Originally a military term embracing details of the transport, quartering and supply of troops, logistical support has broadened to include the provisioning of all hardware and software and the maintenance of these items in support of instruction. In the supplementary figure, 18.0 excludes facilities which are treated separately in 19.0. The decision in 3.3 to perform 15.0, 16.0, 22.0, etc., also triggers an input to *design software* (18.1.1). The basic input is from 3.3 with a secondary input from 17.1. *Perform film/television photography* (18.1.1.1) may range from shooting several 35mm color slides using a copying lens with a SLR camera all the way to producing a 30-minute, 16mm film based on the Sedlik model (29). Subsystem 18.1.1.1 requires graphics to be produced in 18.1.1.2 and sound narration in 18.1.1.3. The training specialist should be conscious of time in this subsystem. Despite the speed of modern photographic laboratories, there are always unexplained and often unjustified delays in shooting and processing unlike those in other industries. Perhaps this is because the software is a creative art form involving a number of different creative skills and sometimes cannot be rushed.

Reproduce consumables (18.1.2) consists of *edit & control* (18.1.2.1), *reproduce* (18.1.2.2) and *bind* (18.1.2.3). These functions can process all kinds of consumables. Several examples are given:

1. *Sound-filmstrip.* Develop storyboard in-house (16.5.3); draw water-color art (18.1.1.2); produce color captions (18.1.1.1); produce sound narration against optical projection of strip (18.1.1.3); synchronize sound/picture, apply pulses (18.1.1.3); edit and control two products (18.1.2.1); reproduce 35mm color internegative, answer prints and final prints (18.1.2.2); reproduce two-track audio cassettes, (18.1.2.2); issue (18.1.3.4); conduct tryout (4.4.1).
2. *Workbook.* Develop workbook concept (15.0); produce lesson involving workbook (4.3); develop detailed workbook with rough sketches (4.8); produce final sketches (18.1.1.2); type text using computer typesetter and dummy pages into camera-ready offset copy (18.1.2.1); offset print (18.1.2.2); bind using perfect edge (18.1.2.3); issue (18.1.3.4); conduct tryout (4.4.1).

3. *Course outline.* Develop outline concept (15.0); produce course outline to teaching point level (4.1.3); synopsize to six pages (4.1.3); design page layout and columns (18.1.1.2); type on 8½ x 11-inch white bond (18.1.2.1); reproduce on electro-static copier (18.1.2.2); collate, punch three holes and staple (18.1.2.3); issue (18.1.3.4); instruct course (5.1).

Handle consumables (18.1.3) consists of *package* (18.1.3.1), *store* (18.1.3.2), *control inventory* (18.1.3.3) and *issue* (18.3.4). Just as in 18.1.2, these functions can process the handling of all kinds of consumables. Several examples are:

1. *Video tape.* Package in electromagnetic-proofed canisters (18.1.3.1); store in security safe (18.1.3.2); control inventory (18.1.3.3); order replacement (18.1.3.3 → 18.1.2.1); issue to training specialist (18.1.3.4).
2. *Textbook.* Package as single copies (18.1.3.1); store in company library (18.1.3.2); control inventory (18.1.3.3); order additional copies (18.1.3.3 → 18.1.2.1); issue to trainees (18.1.3.4).
3. *Flowchart template.* Package as single item (18.1.3.1); store on shelf (18.1.3.2); control inventory (18.1.3.3); order additional items (18.1.3.3 → 18.1.2.1); issue to trainees (18.1.3.4).

Consumables inventories are often controlled by the number of trainees scheduled into courses. The path 13.3 → 18.1.3.3 → 18.1.2.1 represents feedback controlling the output of (18.1.3.3).

Provide hardware support (18.2) consists of *procure* (18.2.1), *maintain* (18.2.2) and *issue* (18.2.3). Technical specifications for the purchasing of equipment result from the inputs of 18.1.2.1 and 17.3 to 18.2.1. Maintenance is often overlooked in 18.2.2, but the success or failure of a program using hardware is a delicate matter. It may involve concepts of Mean-Time-To-Repair (MTTR) and Mean-Time-Between-Failure (MTBF) (32). Experience with maintenance contracts and with poor equipment can be fed back, (18.2.2 → 18.2.1), to modify the specifications for new procurements. For example, a leasing contract for Teletype 33 ASR machines may be terminated with one firm and another firm substituted because of better preventive maintenance (MTTR, MTBF). Also, experience in issuing equipment (18.2.3) and then finding that it is not idiot-proof can result in feedback, (18.2.3 → 18.2.2), safeguarding both equipment and trainee. One illustration comes to mind. Hand-held 8mm optical viewers powered by small drycells were prepared with cells inserted, ready for use. Kept on the shelf, the batteries leaked electrolyte (they were marked "leak-proof!") over the mechanism. As result of 18.2.3 → 18.2.2, batteries were placed in a plastic bag and issued to trainees with instructions for inserting and removing.

When hardware and software are ready, they input for tryout (4.4.1). The major thrust is validation of the software: 4.4.1 → 4.4.2 → 4.5.1 → 4.5.2. However, a tryout at the same time tests the suitability of hardware, revealing maintenance difficulties before the materials are placed into full-scale use in *instruct course* (5.1). Thus the major feedback loops are:

1. Software: 4.5.2 → 3.4 → 3.5 → 3.3 → 4.3.3 → 4.3.1 → 4.1.3, and/or 4.5.2 → 3.4 → 3.5 → 3.3 → 18.1.1 → 18.1.2.1 → 18.1.2.2 → 18.1.2.3 → 18.1.3.1 → 18.1.3.2 → 18.1.3.3 → 18.1.3.4 → 4.4.1
2. Hardware: 4.5.2 → 3.4 → 3.5 → 3.3 → 18.1.1 → 18.1.2.1 → 18.2.1 → 18.2.2 → 18.2.3 → 4.4.1

Most training specialists tend to underplay the role of logistics and wonder why they engage in fire fighting when things fall apart during the program. The key is logistics—the answer is logistical organization and that is where a major expenditure of effort is necessary in the formative stages of a complex project.

Provide Facilities (19.0)

In some industries, training programs are so extensive they occupy entire buildings and, in a few instances, entire campuses.

Recognition of this growing trend to isolate training from the job environment is embodied in 19.0. Facilities include all the physical structures and installed, fixed equipment. The author has been involved in the design of training facilities which had unique electrical power requirements of 400 cycles (33). This called for installation of large motor generator sets in a building set aside for that purpose. These sets were not used in instruction, but the power was. At another time, the author was asked to assist in designing a disaster training facility for heavy rescue teams (34). This was a village of demolished buildings in which "victims" were trapped and trainees were expected to rescue them by digging tunnels, performing high wire escapes, etc. The village had to be built as demolished structures just like a movie set but perfectly safe for trainees to learn without being injured.

Subsystem 19.0 consists of *analyze facility requirements* (19.1), *design/build/modify facility* (19.2) and *provide turnkey operation* (19.3). The analysis in 19.1 might consider a wide range of factors, some of which are:

1. Volume and division of classroom, conference, laboratory and shop space
2. Sanitation and recreation space
3. Storage space
4. Air conditioning and heating
5. Acoustics
6. Access, including aisles, halls and stairwells
7. Parking, including shipping and receiving area
8. Dormitory, including motel/hotel accommodations
9. Transportation, including airport pickup and delivery and car rentals for entertainment purposes
10. Interior decoration

Hopefully, *apply human engineering* (17.0) provides an input to 19.0. An excellent treatment of the human performance engineering field has been published by De Greene which expands upon the subsystems in 17.0. Generally speaking, there is an analysis of man-machine relationships in 17.1. This function has two basic inputs:

1. Feedforward from *quantify parameters* (3.2), which describes the numerical requirements and measurement units for total system performance
2. Information from *formulate learning psychology* (14.0), which describes various learning theories and furnishes research evidence gleaned from experiments.

These inputs influence man-machine analyses performed in 17.1, which next interact with *analyze man-men-facility relationships* (17.2). This subsystem examines the role of individuals and groups within the facility and delineates design requirements to satisfactorily house the program. Inputs from 17.2 and 17.1 to *establish maintenance standards* (17.3) result in maintainability specifications for facilities which have more than routine custodial requirements. In these days of population unrest and mobility, security of records, computer tapes and expensive equipment must be subsumed under 17.3.

When the facility requirements are analyzed and delineated in 19.1, the architects or contractors design, build or modify the facility (19.2). Where this is a large project, it is wise to assign a training specialist as liason so that on-the-spot solutions not provided for in the original drawings can be made without delay. This is shown in the supplementary figure as a feedback path from *manage instruction* (5.3) to *design/build/modify facility* (19.2). It controls decisions made within 19.2 so that the output will meet overall system performance planned in 3.0.

Provide turnkey operation (19.3) is a new concept where a facility is built and turned over to the operating agency in a ready-to-go condition. One need only turn the key in the front door. This presupposes that the builder of the facility also provides operating personnel, recruited, trained and ready to instruct—with all curriculum materials. It is included here as a concept but not necessarily as a recommendation simply because it is

Figure 3-10. Elements from 2.1 in the DB cell of 4.1.1.

very new and largely untried in training environments. There are different degrees of turnkey activity. In any case, the facility is conveyed to the training organization and is managed in 5.3.

Develop Curriculum (4.0)

Regardless of the selection in 3.3 of human-instruction, machine-instruction, computer-assisted instruction or a mix of these methods, it is necessary to *produce course outline* (4.1), *produce tests* (4.2), *produce lesson plans* (4.3), *develop ancillary materials* (4.8), *tryout course* (4.4) and *validate course* (4.5).

Two outputs from 3.3 enter 4.0; 3.3 → 4.1.1 and 3.3 → 4.3.3. The former is the path to be followed when a new course is being planned, while 3.3 → 4.3.3 provides for vernier adjustments to an existing course. When a new course is to be developed, one of the decisions deals with the logical selection and division of the output of 2.1. In Figure 3-10, we see the *DIG/BIA matrix* of 4.1.1 in which the *DB* cell contains *elements* (2.1) directly related and in the basic course. *B, I* and *A* (Basic, Intermediate and Advanced) are arbitrary designations for courses and may be reduced or expanded as each condition warrants. The 4.1.1 decision process is a mental activity.

The logical division of subject matter depends on the real-life productivity needs of the system (6.4.3), the quality needs of the system (6.4.2 and 2.3), the nature of the subject matter (2.1) and the time allowances and budget aspects (3.2) coupled with the subjective constraints or biases of the training specialist. Of these, the most overriding in practice are time and budget. The most economical and realistic decisions will favor elements in DB, DI and DA cells, since they are essential to the performance in 6.4.3.

A course can follow one of three fundamental patterns in 4.1.2 (21):

1. Gestalt
2. Semi-Gestalt
3. Non-Gestalt

Between Gestalt and Non-Gestalt are many shadings of Semi-Gestalt, and in the interest of simplicity it is assumed that these will satisfy all situations. Figure 3-11*a* shows a *Gestalt* model.This follows the principle of analysis:

1. First, the crude whole is presented.
2. Next, the first level of detail is instructed.
3. Then, the second level of detail is instructed.
4. Finally, the last level of superfine detail is presented and limited so as not to lose identity of these parts and relationships to each other.

Throughout the 4.1 phase of preparation for instruction, analysis (identity, relate, separate, limit) is applied. Gestalt is selected and used in 4.1.2 when most or all parts of the subject matter interrelate or somehow interact. This has been called "spiralling," but Gestalt is a more precise descriptor.

At the opposite extreme is the *Non-Gestalt* pattern (4.1.2) depicted in Figure 3-11*b*. The subject matter is cohesive and has a logical, sequential order but each part is studied independently in great detail. When the particular part is completed it is rarely referred to again. The parts are discrete and do not interrelate except serially as a chain of events. The first part to be taught is extremely important since it determines the sequence of those which follow. Sometimes the sequence has a time or chronological basis. In other instances, there is a prerequisite basis where certain

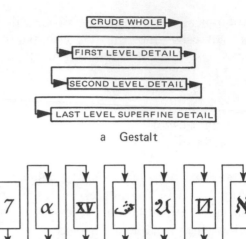

a   Gestalt

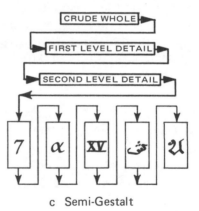

b   Non-Gestalt

c   Semi-Gestalt

Figure 3-11. Patterns selectable in 4.1.2; a, *Gestalt;* b, *Non-Gestalt;* c, *Semi-Gestalt.*

elementary content is reviewed before the actual subject matter can be studied.

Figure 3-11c depicts the *Semi-Gestalt* pattern (4.1.2) where subject matter is taught first as pure Gestalt. However, before the third level of detail is reached, a shift to Non-Gestalt occurs. The parts in the second level of detail which interrelate are next viewed as discrete elements and each is studied in great detail. Once completed, they are not again examined. Obviously, there can be a great many logical combinations and variations of Semi-Gestalt.

In *produce COTP* (4.1.3) the Course Outline to the Teaching Point (COTP) level is produced. It is a mix of method and content. In 3.3 certain methods decisions had been made at the macrolevel. Now, these are implemented at the microlevel in 4.1.3 and deal with such questions as classroom or laboratory, classroom or shop, lecture or conference, etc. Submicrolevel method decisions (such as referring to color 35mm slide projection or halftone in workbook, instructor presenting concept by analysis (whole-to-parts) or concept embedded in problem and guiding trainee through synthesis (parts-to-whole) to solution and establishment of concept) are made in 4.3.1 and not in 4.1.3.

The hierarchy of elements is illustrated in Figure 3-12 (5, 21, 25). The university professor uncomfortable with *program* may substitute *degree-directed program;* the training director will call it *training program;* but both will agree that it consists of courses. When in provide human-instruction (15.0) mode, the *lesson plan* (4.3) almost never breaks down into steps (4.3.3) but simply relies on the memory of the human-instructor. In fact, it is generous to say that human-instruction on a wide scale even has documented teaching points (4.1.3). In *provide machine-instruction* (16.0), the lesson plan (4.3) *must* break down into steps (4.3.3) and Figure 3-12 reveals the extent of this detail. It can be seen that for CAI the step (4.3.3) level is a mid-point rather than end point, since mechanization and synthesization constitute further events in decomposition.

COTP (4.1.3) is the first major phase in organizing and sequentializing content and method. COTP is a chronological sequence of teaching points: 4.1.3 → 4.3.1 → 4.1.3. Structurally, Figure 3-12 shows that the course consists of *units* (or units of instruction as they are frequently called in education) and these consist of *lessons.* In very large programs, another level in the hierarchy is needed and this is inserted as the *subunit* (25). By using subunits, tampering with the sizes of the lessons, teaching point and step are

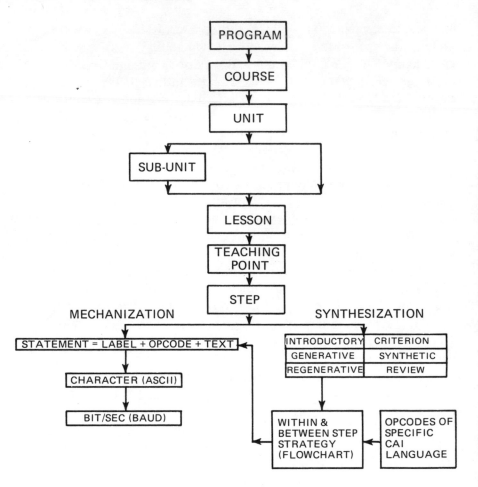

Figure 3-12. Hierarchy of elements.

avoided. It is a convenient mechanism for grappling with courses which are over 100 hours long. From the flow aspect, the supplementary figure reveals that course, unit and subunit, if any, are organized and sequentialized in 4.1.3.

The lesson plan (4.3) consists of teaching points (4.3.1), the *DIG/BIA/OAI* cubical matrix (4.3.2) and steps (4.3.3). Each individual teaching point is created in 4.1.3 and also appears in 4.3.1 as a subset of the lesson plan (4.3). By definition, the *teaching point* is the smallest bit of terminal behavior which it specifies as an outcome of the course. It is a statement of the specific terminal behavior or the smallest bit of knowledge or skill or attitude which the learner is to learn (21). Speaking mathematically, the course content is a Markov chain of teaching points (17). A teaching point is an element of a Markov chain. Teaching points must follow an orderly, step-by-step sequence. The sequence of teaching points is logical and important. One teaching point should be a statement of one fact or one concept—not many. The statement must be precise and unambiguous. This is illustrated with a sequence of typical teaching points:

1.1 Wind is the movement of air.
1.2 Wind can move in any direction.
1.3 Wind can move at any speed.
1.4 Wind can change direction and speed at the same time.

Notice that any resequencing will disturb the logic as seen from a trainee's viewpoint. Figure 3-13 represents a typical COTP (4.1.3). A particularly elementary lesson was selected and each point is stated in unambiguous terms not subject to reader disagreement! Step type is identified in the hierarchy of Figure 3-12 and applies to programmed instruction (16.5.1), teaching machine materials (16.5.2) and computer-assisted instruction (22.2); it does not apply to human-instruction (15.0) as explained above; it may or may not apply to communication aids software (16.5.3) depending on design characteristics and resolution of detail.

There are a number of different step types, but six are adequate for this discussion:

1. Introductory (I). First step to introduce a new teaching point.
2. Generative (G). Generates or causes to be generated and elicited from trainee a word, group of words, diagram or other overt performance.

EDUCATION AND TRAINING CONSULTANTS Co.

COURSE NAME	EXPLORERS OF THE AMERICAS	(Deaf Children)
CURRICULUM DEVELOPER	Robert Edwards	

COURSE CODE 3.1.3.105

DATE 11 Feb 70   TEAM   PAGE 1 OF 6

Unit	Lesson		Teaching Point	Step Type	Step Nos.
1.4 Christopher Columbus' Four Voyages to the Americas	1.4.1 Christopher Columbus and Events That Affected His Life	1.4.1.1	At the time of Christopher Columbus' birth, many people believed the earth was flat	I	
		1.4.1.2	Other people were sure the earth was round	I	
		1.4.1.3	People wanted to find a new route to Asia	IG	
		1.4.1.4	They wished to obtain silks, spices and gold	IGS	
		1.4.1.5	Some people thought an all-water route to Asia could be found	IG	
		1.4.1.6	These people believed this route could be found by crossing the Atlantic Ocean	IG	
		1.4.1.7	These people did not know there were two large bodies of land between Europe and Asia	IG	
		1.4.1.8	Most sailors were afraid to sail across an unknown sea	IG	
		1.4.1.9	Christopher Columbus' home was in Genoa, Italy	IG	
		1.4.1.10	Genoa was a seaport	IGS	
		1.4.1.11	Columbus often talked with sailors about their voyages	IG	
		1.4.1.12	He would then plot their courses on his home-made maps	IGS	
		1.4.1.13	He became a sailor when he was 14 years old	IG	
		1.4.1.14	He sailed to Greece, Asia Minor, England, Ireland, Iceland and the west coast of Africa	IG	
		1.4.1.15	Columbus went to Portugal to study mapmaking	IG	
		1.4.1.16	He became a captain for the Portugese	IG	
		1.4.1.17	He made a good living as a sea captain	IGS	
		1.4.1.18	He studied books and maps	IG	
		1.4.1.19	He believed it was about 2,400 miles from Europe to Asia	IGS	
		1.4.1.20	Columbus had a plan to find a water route from Europe to Asia	IG	
		1.4.1.21	He believed he could travel west and reach Asia	IG	
	1.4.2 Christopher Columbus' First Voyage to the Americas	1.4.2.1	Columbus needed money for men and ships	IG	
		1.4.2.2	He asked the King of Portugal for money	IG	
		1.4.2.3	The King of Portugal did not give him the money	IG	
		1.4.2.4	He then asked the King of Spain for money	IS	

Figure 3-13. A typical COTP.

3. Regenerative (R). Regenerates or causes to be regenerated and elicited from trainee essentially the same word, group of words, diagram or other overt performance by rearranging the stimulus.
4. Synthetic (S). Relates old teaching point to new teaching point and combines them, thereby eliciting a new, overt performance based partly on an existing behavior.
5. Criterion (C). Ascertains if trainee remembers and understands a previously presented teaching point.
6. Review (Re). Provides a previously presented step with virtually no change.

Step type is entered on the COTP form first as an estimate. When the steps are actually written in 4.3.1, corrections may be entered in the COTP step type column. This provides rough cut decisions following the feedback path: 4.3.3 → 4.3.1 → 4.1.3, thereby controlling outputs of 4.1.3 and 4.3.1. *Step numbers* are entered after step type has been identified and have utility only for programmed instruction and CAI. There are continual interactions as the COTP evolves or is synthesized into a final draft: 4.1.3 ⇌ 4.3.1 ⇌ 4.3.2 → 4.3.3 → 4.3.1. In essence, *sequentializing* requires the creation of several new teaching points which bridge a gap between two existing teaching points so the learner will learn. Analysis at this microlevel reveals there is a gap. Synthesis produces several new teaching points, identifies, relates and combines these with the two existing teaching points. A new whole has been produced which is larger than the previous whole. It is larger than merely the sum of the points taken as separate parts. Thus, *synergism* operates to produce a new whole which is greater than the same parts viewed as a discretum. In most instances, it is the structure or organization or logic of the subject matter which dictates the general logic of the sequence of teaching points. Badly structured subject matter invariably produces illogical sequences of teaching points.

Create pre/post/other tests (4.2.1) is derived directly from *produce COTP* (4.1.3). A polished draft is produced when the

COTP is prepared. Only a few changes are expected in 4.2.1 as a result of *tryout* (4.4) and *validate* (9.5). These tend to clarify or otherwise reword a question rather than tamper with the statistics in the *gain* (4.2.2) function. Any attempts to skew these results by sweetening any of the tests (4.2.1) should be avoided. To put it another way, the COTP (4.1.3) with its teaching points (4.3.1) is the only source of test items. Hopefully, these will be performance test items rather than simply the paper-and-pencil variety. The output of (4.2.2) enters (4.5.1) and determines the validity after tryout.

The transition from a single teaching point (4.3.1) to a series of steps (4.3.3) which will optimally elicit prescribed trainee behavior is difficult. That is, consistent instructional programming of high quality at a low time expenditure is not a simple matter. Programming is a mix of art and science and it separates the men from the boys (and the women from the girls!).

The *DIG/BIA matrix* (4.1.1) is a two-dimensional square. It is expanded to a three-dimensional cube by adding a *Z*-axis consisting of *OAI*: *O*bjects, *A*ctions and *I*nformation. The nine cells (possibilities) in 4.1.1 increase to 27 cells in (4.3.2). Once again, this decision-making is a mental process which assists in converting one teaching point to a series of steps. If the cell is *DBO*, we are dealing with a *d*irectly related teaching point involving a physical *o*bject in the *B*asic course. The emphasis will be on object analysis or object synthesis but only of essential features as revealed in 2.2 and at the level of proficiency quantified in 2.3.

It has been shown in Figure 3-12 that each teaching point has one or more steps. The entry on the COTP was an estimate first of the number of such steps and second of the kind of steps.

Now in 4.3.3, the instructional programmer writes the series of steps according to his prior number-kind estimate. The quality of this estimate will be a function of and vary directly with his experience. In Figure 3-12, synthesization is the process of identifying, relating and combining steps (4.3.3) and it must consider several aspects of method. A step is treated both from mechanization and synthesization viewpoints. *Mechanization* is the *physical* breaking down, while *synthesization* is the *conceptual* building up. The six kinds of steps apply mainly to programmed instruction and CAI in the tutorial mode, which emphasizes conversation. The rationale for sequencing, for example I-G-S of 1.4.1.4 in Figure 3-13, is to elicit appropriate trainee performance and is based on a particular theory or set of theories in *formulate learning psychology* (14.0). The path is $14.0 \rightarrow 3.3 \rightarrow 4.1.1 \rightarrow 4.1.2 \rightarrow 4.1.3 \rightarrow 4.3.1 \rightarrow 4.3.2 \rightarrow 4.3.3$. The machine screw path provides vernier corrections consisting of feedback: $3.3 \rightarrow 4.3.3$. The limit switch at the head of the screw entering 4.3.3 feeds a signal back to 2.1 if rewriting steps in 4.3.3 fails to elicit the desired behavior. The feedback signal to the JA-HAA reexamines the element, its level of proficiency and relatedness to real-life. Here are several alternatives:

1. Relatedness = D, LOP = 10; the element is essential; $4.3.3 \rightarrow 2.1 \rightarrow 3.1 \rightarrow 3.2 \rightarrow 3.3 \rightarrow 4.1.1 \rightarrow \ldots 4.3.3$. Programmer must rework until behavior is elicited. This is often accomplished by increasing and improving the sequencing of step types in 4.3.3.
2. Relatedness = I or G, LOP $\leqslant$ 9; the element is not essential; $4.3.3 \rightarrow 2.1 \rightarrow 3.1 \rightarrow 3.2 \rightarrow 3.3 \rightarrow 4.1.1 \rightarrow \ldots 4.3.3$. Programmer does not have the "absolute" requirement and reworks until the less critical behavior is elicited.

A step cannot be converted conceptually into smaller parts or it will lose its identity. Mechanization is the physical breakdown used only in the CAI path: $22.3 \rightarrow 23.2$ or similar $\rightarrow 23.7 \rightarrow 23.8 \rightarrow 23.12$.

When the lesson plans (4.3), hardware (18.2.3), software (18.1.3.4), ancillary materials (4.8), tests (4.2.1) and CAI programs (23.12), or any combination of these input (4.4.1) the tryout is conducted. The trainee target population is sampled, and a percentage is selected to tryout the instruction. An analysis is performed in 4.4.2. Here are some alternatives:

1. Human-instruction; analyze test data.
2. Machine-instruction; analyze specific responses to steps; analyze test data.

3. Computer-assisted instruction; analyze patterns of responses; analyze specific responses to steps; analyze test data.

In *validate course* (4.5) decisions are made concerning the effectiveness of the instruction in quantitative terms. The major input is from *compute gain* (4.2.2). If 3.2 has stated earlier that $X\%$ of trainees must achieve a gain of Y, the validation decision rests on these measurements. Some of the alternatives in 4.5.2 are:

1. Accept: $4.5.2 \rightarrow 5.1, 4.5.2 \rightarrow 4.7 \rightarrow 5.3 \rightarrow 5.1$
2. Reject: $4.5.2 \rightarrow 3.4 \rightarrow 3.5 \rightarrow 21.0$, where the CAI processor is improved
3. Reject: $4.5.2 \rightarrow 3.4 \rightarrow 3.5 \rightarrow 3.3 \rightarrow 4.3.3 \rightarrow 4.3.1$, where several steps are rewritten or teaching points reworked
4. Reject: $4.5.2 \rightarrow 3.4 \rightarrow 3.5 \rightarrow 3.3 \rightarrow 4.1.1 \rightarrow 4.1.2 \rightarrow 4.1.3$, where the COTP is reworked
5. Reject: $4.5.2 \rightarrow 3.4 \rightarrow 3.5 \rightarrow 3.3 \rightarrow 18.1.1$, where software is reworked
6. Reject: $4.5.2 \rightarrow 3.4 \rightarrow 3.5 \rightarrow 3.3 \rightarrow 15.0$, where instructor is replaced with different instructor
7. Reject: $4.5.2 \rightarrow 3.4 \rightarrow 3.5 \rightarrow 3.3$, where different method is selected
8. Reject: $4.5.2 \rightarrow 3.4 \rightarrow 3.5 \rightarrow 4.3.3 \rightarrow 4.6 \rightarrow 2.1$, where JA-HAA is reviewed

To summarize, if rejection occurs in 4.5.2 the immediate task is troubleshooting and the training specialist must localize the malfunction, then correct it. Where to look in the system and what to do depends on the symptoms detected in 4.5.1. In this regard, it is like a malfunction in a watch, car, television set or any complex machinery. Troubleshooting demands a competent troubleshooter or some clown will put more trouble into the system.

When a course is validated, it is customary in large, decentralized organizations to produce an administrative factors guide (4.7). This is documentation in the form of a manual describing the program, how it was validated, tryout data, etc. A good example is Reference 35. It permits training managers and administrators in operating divisions to examine the terminal behavioral objectives of the course and the validation data before actually assigning employees to take it.

The signal error function (4.6) receives inputs from *evaluate course* (5.2), *instruct course* (5.1), *filter distortion* (24.0) and *supervisor* (6.4.1). These are the various states:

1. From evaluate course (5.2); based on instruction (5.1) and utilized to correct errors detected by training specialist by time course ends, corrective action would be 4.6 → 4.1.1 → . . . 4.3.3 or 4.6 → 2.1 → . . . 4.3.3 depending on severity of errors.
2. From instruct course (5.1); based on instruction (5.1), trainee is invited to comment on effectiveness at time course ends; replies are filtered for distortion and input (4.6); corrective action usually is 4.6 → 4.1.1 → . . . 4.3.3 although it is possible to follow 4.6 → 2.1 → . . . 4.3.3 if the errors are severe.
3. From supervisor (6.4.1); based on supervisor perception of graduate trainee performance (6.4.3 → 6.4.1. Supervisor corrects errors performed by graduate immediately by re-instructing him, 6.4.1 → 6.4.3). This feedback controls the performance output of 6.4.3 and, consequently, the outputs of 6.1.1, 6.1.2 and 6.1.3. He relies on standards in 6.4.2 which input 6.4.1 to judge the graduates performance. Having corrected these errors immediately, he also sends a feedback signal to 4.6 so corrections can be made in the course. This controls the errors in the course so graduates of forthcoming courses will not perform below standard on the job.

Conduct Program (5.0)

To instruct course (5.1) the training organization requires these kinds of inputs:

1. Humans
    a. Trainees who were previously recruited (13.2) and scheduled (13.3) into 5.1, based on system parameters (3.2) and selection criteria (13.1)
    b. Instructors who were previsouly selected (15.0) and assigned to (5.1)
    c. Managers who were previously selected (5.3) and assigned to (5.1)
    d. Maintenance personnel who were previously selected (19.3) and assigned to (5.1)
2. Course method and content
    a. Lesson plans (4.3); ancillary materials (4.8); and other software (23.12), (18.1.3), (16.3)
    b. Tests and similar criterion measurement instruments (4.2)
3. Hardware (18.2.3)
4. Facilities (19.3 → 5.3)

While the course is being conducted, evaluations through the use of tests can immediately detect and correct content malfunctions by applying feedback (5.2 → 5.1). Method malfunctions are immediately controlled by feedback (5.2 → 5.3). These feedback loops correct the course while it is operating in contrast with others which correct subsequent courses.

In some instances, the program is excellent but the trainees being assigned are not qualified to attend. This may be a gradual trend rather than an immediate problem. If it is gradual, then a feedback from evaluation (5.2) to recruitment (13.2) signals a review of recruiting procedure and criteria thereby controlling that function. In recruiting it might be concluded that the selection criteria should be changed and a feedback from 13.2 to 13.1 would control this.

Formulate Learning Psychology (14.0)

For many years, Hilgard has attempted to bridge the gap between learning theories and classroom applications. About ten years ago, he decried, " . . . present bridges between the laboratory and classroom are so flimsy . . . " (36). At the time, he was asked to propose principles for practitioners and said, "I believe these principles to be in large part acceptable to all parties, and the affiliation with one or another source is a matter of emphasis rather than controversy" (36). The 14.0 subsystem is patterned after the Hilgard synthesis of the following principles:

14.1 Principles Emphasized by S-R Theory

14.1.1 The trainee should be *active,* rather than a passive listener or viewer. S-R theory emphasizes the significance of the learner's responses. "Learning by doing" is still an acceptable slogan.

14.1.2 *Frequency* of repetition is important in acquiring skill and in bringing enough overlearning to guarantee retention.

14.1.3 *Reinforcement* is important; repetition should be under arrangements in which correct responses are rewarded. It is generally found that positive reinforcements (rewards) are to be preferred to negative reinforcements (punishments).

14.1.4 *Generalization* and *discrimination* suggest the importance of practice in varied contexts so that learning will become (or remain) appropriate to a wider (or more restricted) range of stimuli.

14.1.5 *Drive* conditions are important in learning, but all personal-social motives do not conform to the drive interpretations of Hull and Spence. Anxiety appears to act as a drive should, but achievement motivation apparently does not.

14.1.6 *Knowledge of results* is the notion that the trainee tries something provisionally and confirms his attempt by its consequences.

14.1.7 *Conflicts* attendant upon generalization and discrimination have consequences that may be unintended by the person attempting to manage the learning.

14.2 Principles Emphasized by Cognitive Theory

14.2.1 A learning problem should be so structured and presented that the essential relationships are open to the inspection of the trainee. Thus, *perceptual aspects* of the problem (figure-ground relations, directional signs, what leads to what) represent important features.

14.2.2 The direction from simple to complex is from simplified wholes to more complex wholes.

14.2.3 Learning with *understanding* is more permanent and more transferable than rote learning or learning by formula. This generalization gives point to the similar emphasis within S-R theory on meaningfulness as a factor making for ease of acquisition and recall.

14.2.4 *Cognitive feedback* establishes probabilities and (in some cases at least) is an appropriate explanation of effective reinforcement.

14.2.5 *Goal-setting* by the trainee is important as motivation for learning, and his successes and failures are determiners of how he sets his future goals.

14.3 Principles Emphasized by Personality Theory

14.3.1 The trainee's *abilities* are important, and provisions have to be made for the slower and more rapid learners.

14.3.2 Some abilities are a matter of physiological and social *development* and knowledge of development should be related to the demands made on the trainee.

14.3.3 and 14.3.4 Personality is a *social product.* Hence, it is important to be aware of the culture and subculture as they are relevant to what and how the trainee can learn.

14.3.5 *Anxiety level* appears important in determining the beneficial or detrimental influence of praise and blame. The generalization appears justified that with some kinds of tasks high-anxious learners perform better if *not* reminded as to how well (or poorly) they are doing, while low-anxious learners do better if they *are* interrupted with comments on their progress.

14.3.6 The same objective situation may tap *appropriate motives* of one learner and not of another. The *organization* of motives and values in the individual is relevant. Some long-range goals affect short-range activities.

14.3.7 The *group atmosphere* of learning (competition versus cooperation, authoritarianism versus democracy, individual isolation versus group identification) will affect satisfaction in learning as well as the products of learning.

Within 14.1, 14.2 and 14.3 the various principles appear to be related, as shown in the supplementary figure, but the exact relationship is either unclear or tenuous. In the decade following Hilgard's statements, very little has been disclosed to alter these principles. Hopefully, Knowles in his forthcoming text on *Learning Theory and Adults,* for this series will add to the principles set forth by Hilgard (37).

Thus, we see formulate learning psychology (14.0) as a lumped whole consisting of many theories and principles which form the basis for decisions in:

1. Select methods (3.3)
2. Provide human instruction (15.0)
3. Establish step format (16.6)
4. Analyze man-machine relationships (17.1)

Let us assume that instructional programming is performed by a person with instructing experience. Silvern contends that one prerequisite is". . . five years of active classroom, shop, or laboratory teaching experience at a grade or age level and in subject-matter area to be programmed. Tutoring or coaching experience on a full-time basis may be substituted" (5). Then it is clear that 14.0 enters 4.3.1 via 15.0 and is the basis for lesson plans to the teaching point or step level for every method of instruction.

Decisions Involving Synthesis

Less than 1% of the training programs in the typical company involve occupations or tasks which have never existed before. The issue of lead-time in large systems has been explored briefly in Chapter 2. If one projects that *invent & innovate* (10.0) will increase its output in the next 30 years, it is conceivable the 1% might rise to 5% or higher. It must be understood that 10.0 is where *all* new ideas are generated. Not just transistors, SSTs, heart transplant surgery or domestic satellites, but also social prohibition on smoking, 18-year old vote, anti-pollution, prison reform, drug control, new relations with mainland China and similar social inventions and innovations. The output from 10.0 is increasing at an almost frightening rate—much of it is subtle to the extent it is upon us without warning. Politics is not immune; Republicans turned Democrat, Democrat platforms implemented by Republican administrators. There are, in fact, two kinds of social revolution in the United States. One is evolutionary, quickened by unrest and dissatisfaction of younger citizens. Another is revolutionary, spear-headed by militants who are purveyors of pain. Both will bring major change to society in general—and to business, industry and government in particular. All of this increases output from 10.0 to:

1. New performance (6.4.3): new products and new services (6.1)
2. New standards (6.4.1): new work relationships and new performance expectations

The role of supervisor (6.4.1) is seriously affected by these new inputs. Patton (38) contends:

> ... the foreman of today is subject to additional pressures which did not exist in prewar America. Among these are the unprecedented power wielded by labor unions, workers' rights and a grievance procedure under which a foreman who disciplines a worker may frequently be called upon to defend his actions, not only to the union, but to top management. As an end result the caliber and makeup of the first-line supervisor has changed for the worse.

Setting aside Patton's recommendation for first-line supervisory training, which is only one of a thousand programs, it is important to consider how to train the non-supervisory employee who constitutes the major training load in occupations and tasks just being born in 10.0. Lead-time increases as job content grows more complex and we must implement the input of 10.0 to *create elements* (1.1) as quickly as possible after detecting 10.0 → 6.4.2 and 10.0 → 6.4.3.

Synthesis is a synonym for creativity and inductive reasoning. When one is highly skilled in synthesis, he is an inventive, innovative person. Job synthesis requires intimate coordination of inventor or his surrogate and the training specialist. Elements are created in 1.1 at the same time they are created in 10.0. The relationship with reality—ulitmate job reality—must be established in 1.2. A quantification of levels of proficiencey occurs in 1.3. The first document should be prominently stamped "Rough Draft" and submitted to another group for analysis. The JA-HAA group (2.1) can examine the output of 1.1 objectively and ask questions of the JS-HAS group, 2.1 → 1.1, which feeds back and forces 1.1 to dig in and create even more information. The interaction 1.1 ⇌ 2.1 requires 1.1 to synthesize continuously and 2.1 to follow by analyzing what is being synthesized. Thus, a better product will result as 2.1 outputs to 3.1.

Just as one must *perform basic analysis* (12.0) for 2.0, there is a requirement to *perform basic synthesis* (11.0) for 1.0. In fact, 1.0 cannot be accomplished without 11.0—the skills are prerequisite.

The model for synthesis appears in the supplementary figure and generally consists of:

1. Identify (11.1) one part.
2. Identify (11.1) a second part.
3. Relate (11.2) the parts to each other, if possible.
   a. If there is a relation, then 11.2 → 11.3.
   b. If there is no relation, then 11.2 → 11.1 a 3rd. . . nth part.
4. Combine (11.3) the parts and form a new whole.
   a. If there is good combination, then determine 11.4.
   b. If the combination produces the largest whole, then 11.3 → 11.4.

    c. If the combination is good but not the largest whole, then
11.3 → 11.1.

    d. If the combination is bad, then reidentify 11.1.

5. Limit 11.4 by halting.

    a. If there is another whole suddenly discovered which might
be synthesized, then 11.4 → 11.1.

    b. If this is the largest whole then, 11.4 → 1.0.

In comparison with analysis, synthesis requires many more decisions and there are alternatives or branches for each decision. This is frequently referred to as the "if-then" syndrome. The steps 1-5, above, are not absolute but constitute a general narrative model; the general flowchart model is 11.0. When there are a very large number of decisions or tests for a yes-no and a very small number of action functions, the flowchart model takes on a style characteristic of computer flowchart models. It is only in this way that machine-simulation or, more precisely, computer simulation can be accomplished.

In this chapter, we have examined a flowchart model with 164 subsystems and hundreds of signal path relationships. There were 41 feedback paths alone. Yet, it is easy in retrospect to see that any one subsystem at the lowest level of detail could have been further refined and modeled to four, five, six. . . ten levels! Every subsystem was relevant to the training process—every signal path represented a critical relationship. Even so, the model is incomplete because it is a *general* model. It can accomodate many problems and produce solutions but it cannot produce answers to *all* training problems. It remains for the reader to begin with the model in the supplementary figure, use it, and refine it to satisfy his more specific needs.

4-simulation

Concept of Simulation

The term *anasynthesis* is defined as a process consisting of four major parts: analysis, synthesis, modeling and simulation (1). These often follow in sequential order:

1. *Analysis* is performed on existing information; in this book, an analysis was made of all important requirements for a total or fully integrated training program. Twenty-three such requirements were analyzed.
2. *Synthesis* is performed to combine unrelated elements and relationships into a new whole; in this book, synthesis produced a new whole of 23 elements having 44 relationships of which 12 were feedback.
3. *Models* are constructed which can predict effectiveness without actual implementation of the system; in this book, modeling produced flowchart analogs of the
   a. 23 systems depicted in Figure 2-8 at the major level of detail
   b. 164 subsystems in the supplementary figure at the third level of detail

A few mathematical models were presented in Chapter 3 but these are considered trivial in terms of what can be accomplished in a major thrust at quantification.

4. *Simulation* is performed which reveals alternative solutions or at least a best solution.

This chapter is devoted to the fourth stage in the anasynthesis process—simulation. Simulation may be viewed as having two purposes:

1. To test the model and debug it until it seems to have a very high correspondence with reality. Problems are input, processed and reprocessed iteratively until all elements and their interrelationships are identified and appear embedded in the appropriate place. Then, the model is frozen; i.e., its design is fixed.
2. The frozen model, having a theoretical one-to-one correspondence, or a fidelity of 100% (high fidelity) with respect to real-life, is used as a problem-solving device.

In this chapter, it is assumed that the flowchart model in the supplementary figure has been tested, debugged and finally frozen. Therefore, we will take a typical problem which the average training specialist might solve in his real-life environment and process it through the model in this figure.

Problem: Train Factory Employees in Fire-Fighting Techniques

Scenario

The company has 22,000 employees in four operating divisions located in New York, Tennessee, Kansas, California and a corporate office in Florida, where the fifth operating division is currently being constructed. The product line is essentially furniture but is diversifying into various wood products. Over 20,000 of the employees are engaged in manufacturing which involves woods, plastics, paints, glues and similar highly combustible materials. While automatic and semiautomatic equipment designed for the ultimate in safety is used, there is always the danger of explosion and fire. Each plant has a full-time fire department on the premises and adequate deposits of emergency equipment placed strategically throughout the buildings. Resources of the community fire department are also available when an alarm is received. However, employees come from a lower middle class culture where property value does not rank high. Recently, a fire broke out in the up-state New York plant and these conclusions were reached by management after a $3,000,000 loss:

1. Employees who tried to assist in fire fighting did not have any knowledge of emergency fire-fighting equipment and were totally ineffective.
2. Some employees stood by, enjoying the confusion and destruction as spectators, not understanding that the loss could shut the plant down and affect their incomes. Endangering their own lives, they also affected the morale of the employees who attempted to suppress the flames.

The training organization has been directed to correct the situation expeditiously and effectively. Management criteria for "effectively" are analyses conducted by industrial engineers and fire officers of all fires after the training has been initiated. This is to determine if the employees engaged in fire-fighting function properly and if spectator activity has diminished. While fires cannot be predicted or totally prevented, employee behavior during fires can be observed and analyzed *a posteriori*. Management desires to have every plant employee meet a minimum standard; employees can be scheduled into training on a demand basis; training shall reoccur periodically to maintain employee skills; employees are to remain on the payroll during training to conform with the law.

Simulation

6.1 Confer with plant fire chiefs and plant industrial engineers. Chiefs believe each employee should have about four hours of individual instruction plus two hours of team instruction using presently deployed equipment. Engineers agree with chiefs but also point out that they are planning to contract for a large number of portable turbine pumps and expect delivery to begin in five months. These pumps are different from any equipment now in use and will be supplemental.

6.4.1 Confer with first-line supervisors selected from departments having fire losses within the last 10 years. Supervisors agree on fire-fighting instruction but they also agree that part of problem is attitudinal. During an emergency, there isn't any time or desire to apply disciplinary measures to spectators. An employee running a wood lathe is not employed to operate a fire hose and the employment contracts do not specify that emergency fire-fighting is a duty. It is an unwritten understanding that employees fight fire until help arrives or until they judge their fire-fighting actions no longer feasible due to the danger. Supervisors believe that some instruction is necessary to explain how loss due to fire can result in unemployment, etc.

6.1 → 7.1 → 7.2 Obtain *existing* fire-fighting knowledge and skill elements from chiefs. Filter questionable elements (7.2) which deal with knot-tying back to chiefs, 7.2 → 6.4.1 → 6.1, to obtain clarification of which single knot can be learned to cover all reasonable circumstances. Feed output to 2.1 for JA-HAA of "emergency fire fighter."

10.0 → 6.4.2 → 6.4.1 → 6.1 Obtain *new* knowledge and skill elements from industrial engineers regarding portable turbine pumps using manufacturer manuals, locations of equipment and how engineers anticipate using them.

10.0 → 1.1 Feed portable pump output to 1.1 for JS-HSS treatment dealing with innovative equipment. Utilize 11.0.

1.1 → 2.1 Analyze *new* knowledge and skills just synthesized and integrate them with existing knowledge and skills. Utilize 12.0

2.1 → 1.1 Control the output of 1.1 if it becomes too large and out of proportion to that of 2.1.

2.1 → 3.1 Establish operating parameters:
1. Hours/trainee—individual instruction
2. Hours/trainee—team instruction
3. Hours/trainee—periodic refresher instruction
4. Total number of employees to be trained
5. Trainees/month completing individual instruction
6. Trainees/month completing team instruction
7. Elapsed time for taking refresher instruction (months)
8. Augmentation of fire instructor staff (instructors)
9. Budget allocation for instructional materials; practice fire-fighting equipment, consumable materials and maintenance of equipment

3.1 → 3.2 Quantify parameters:
1. Hours/trainee for individual instruction = 4
2. Hours/trainee for team instruction = 2
3. Hours/trainee after first year for periodic team refresher instruction = 2
4. Total number employees in company = 22,000
5. Total number employees to be trained = 20,000
6. Trainees/month to complete four-hour course = 2000
7. Trainees/month to complete two-hour course; beginning month two = 2000
8. Elapsed time for taking refresher instruction (months) = 12
9. Augmentation of fire instructor staff; each fire officer is capable of instruction while on duty

   a. Fire Instructor Staff

Plant Loacation	Employees to be Trained	Present Staff Fire Officers
California	5000	30
Florida	0	0
Kansas	5000	30
New York	5000	30
Tennessee	5000	30

b. Each plant has two engine companies; each company has a five officer crew; two companies are on duty for eight hours or one shift; there are three shifts; while on duty, all fire officers are currently engaged in fire inspections; one company can be assigned to provide instruction while the other company makes routine daily inspections—this will reduce daily inspections 50% but will make trained employees available. Thus, the *present* fire officer force can furnish five instructors per day, per shift, per plant.

c. Assuming zero absences by instructors from instructional duties:
1. Five instructors can furnish 5 x 8 or 40 hours of instruction per day, per shift, per plant.
2. This equals 40 x 3 or 120 hours of instruction per 24-hour day per plant.
3. Five such days (one work week) provide 120 x 5 or 600 instructing hours per week per plant.
4. With 50 work weeks each year, 600 x 50 or 30,000 instructing hours yearly would be available under ideal conditions in each plant excluding Florida.

d. Assuming zero employee turnover and availability on demand without delays, excuses or absences:
1. 5000 employees would receive 5000 x 6 or 30,000 hours of instruction.
2. If classes consist of several rather than single trainees, or if machine-instruction is used, there will be an ample number of instructors.

e. Alternatives if instructor shortage develops:
1. Increase class size; trainee groups will still be very small.
2. Automate some part of individual instruction; this is easily done since it is repetitious.
3. Recruit and hire a few fire officers in each plant; cost would be $10,000 per man-year.

10. Budget allocation for instructional materials, etc.
    a. For classroom type materials, estimate $1.00 per train-
        ee 1 x 20,000 or $20,000; each plant would receive
        $5000 for the first year.
    b. For equipment, consumable materials, replacement, es-
        timate $2.00 per trainee 2 x 20,000 or $40,000; each
        plant would receive $10,000 for the first year.
    c. Second-year costs will depend on employee turnover
        rate, situation in Florida plant, experience in fire
        fighting, fire frequency per plant, and so forth.

3.2 → 13.1 Set screening criteria
    1. All hourly and salaried employees physically assigned to
        buildings in which manufacturing, storing or shipping are
        conducted are eligible.
    2. Employees who frequent such areas at least five hours a
        week but who are regularly assigned elsewhere (trucking,
        design, sales) are eligible.
    3. Paygrades from X-1 (entry occupations) to X-10 (junior
        executive management) are to be trained.

13.1 → 13.2 Recruit
    1. Supervisors are consulted to determine how many employ-
        ees can be spared, at which periods during each shift, at
        what periods during each month and when time off or
        vacations are scheduled in their departments.
    2. Union representatives are consulted to establish the sup-
        port of organized labor and bargaining units for the
        training program.
    3. There is strong feeling by union stewards that paygrade
        X-10 (junior executive management) is too low as a cutoff;
        training is to preserve the physical plant and possibly save
        lives; therefore reconsider 13.1 and raise X-10 to include
        X-16 (senior executive management).

13.2 → 13.1 Feedback to 13.1 produces new paygrade range of
    X-1 to X-16, inclusive.

13.1 → 13.2 Internal recruitment authorized by issuance of cor-
    porate policy bulletin (CPB) posted on all official bulletin
    boards and distributed to all supervisors.

13.2 → 13.3 First line supervisors prepare to schedule employees as trainees; management prepares to schedule supervisors.

3.2 → 3.3 Methods selected; program consists of courses based on input from 14.0 (notably 14.1.1, 14.3.3 and 14.3.7):

1. Introductory Course (IC)—four hours
    a. Programmed instruction, two hours; studied on job near work station, in company cafeteria or in company library.
    b. Classroom and fire ground instruction, two hours; at company fire station.
2. Team Course (TC)—two hours
    a. Fire ground instruction, two hours; at company fire station.
    b. Trainees must work in same department and are scheduled in 13.3 as a team.
3. Team Refresher Course (TRC)—two hours
    a. Fire ground instruction, two hours; at company fire station.
    b. Trainees form into same teams as for TC; replacements due to transfer, termination or promotion are made by supervision in 13.3.
    c. Begins 12 months after start of program; eligibility is completion of IC and TC.

3.3 → 15.0 Training director and fire chiefs decide

1. Fire officer shall receive a short intensive instructor training course, 14.0 → 15.0 (ITC), so all will be eligible to instruct; this will allow shift rotation, vacations and illness without disrupting program; cost estimate is $50 per officer; 50 x 120 or $6000 will be charged to the training organization's budget.
2. Officers who do exceptionally well in ITC will be assigned by chiefs as instructors; ITC is to be conducted in each plant by the training organization and officer performance is rated by training specialists; ITC is decentralized.
3. One officer from each of the four plants will be assigned by fire chiefs as members of the Fire Training Advisory

Committee, with the title of Fire Training Officer, for developing curriculum materials, selecting subject matter and coordinating intra-plant and inter-plant activities. He will also direct the training program and manage the fireground.

3.3 → 16.1 Training specialists and the Fire Training Advisory Committee examine existing commercial courses and decide that, due to requirements for improving employee attitudes at disasters and the unique character of plant fire-fighting equipment, the best solution is to *make* the programmed instruction component of IC. This is expected to consume two hours per trainee.

16.1 → 16.4.2 The training organization is decentralized, having an office in each of the four operating facilities. Since the New York plant recently had a $3,000,000 loss and several training specialists there are proficient in writing programmed instructional materials, the decision is to have New York produce a two-hour training package in-house. This will be used in California, Kansas, New York and Tennessee. It may also be used in Florida but this will depend on actual experiences in conducting IC.

16.4.2 → 16.5 There is uncertainty regarding the design of the materials for the two-hour programmed instruction component of IC. The issue is one of cost. The text in 16.5.1 would be expendable and a printing of 20,000 would be necessary. The employees could keep the text but it is unlikely that they would ever refer to it unless it contained highly useful information. A machine in 16.5.2 would require a great initial capital outlay and would slow down training unless many machines were purchased and deployed in the libraries, cafeterias and work stations. A typical filmstrip, branching machine costs $1000 without the software. The training organization's current budget does not allow any such purchases and the $20,000 estimate in 3.3 would permit only five machines per plant. Of course, these machines could be leased for one year until the major effort has been completed, but high-speed maintenance is not available in all the states involved. Also, several training specialists favor trainee constructed response over multiple-choice branching and this would negate a 16.5.2 decision. The training period is too short to warrant a teaching machine procurement.

The combined task force of training specialists and the Fire Training Advisory Committee reason that:

1. A typical trainee can work through 120 steps in two hours.
2. One physical page of 8 1/2 x 11-inch paper contains four steps, printed on one side.
3. A text would consist of about 15 pages, printed on two sides, plus cover and mask.
4. Cost estimate (materials):

1,000	covers and masks, loose-leaf, reusable, @ $2.00	$4000
20,000	15-page text, loose-leaf, printed on 2 sides, collated and stapled	$4500

5. Cost estimate (training specialists): $1068
   178 production hours to produce
   two hours of programmed text @ $6.00/Hour
6. Total cost without considering trainee salaries
   during tryout (4.4.1) $9568
7. Constructed response is based on operant conditioning techniques, 14.1; also 14.2.3 would be involved in recall rather than simply recognition in a multiple-choice response format.
8. Introduction of new turbine pumps by the industrial engineers can easily be accommodated in a printed format but with less facility and at higher cost in a teaching machine filmstrip format.

The decision is reached in 16.5.1 to design a programmed text.

16.5.1 Programmed text is designed by training specialists using specific terminal objectives based on 2.1; each trainee completing IC of two-hour duration shall:

1. Understand the consequences to himself and his family if his employment is terminated due to fire loss
2. Be able to compute the financial loss suffered in terms of hospital, surgical, medical and life insurance coverage by himself and family
3. Be able to compute the long-term financial loss from cancellation of annuity and stock-sharing plan, based on his present age
4. Be able to compute the immediate financial loss resulting from a mandatory transfer to another plant; selling home, uprooting children from school, etc.
5. Be able to trade-off existing company benefits, plus benefits planning through union bargaining activity, against termination of employment due to fire disaster
6. Be able to select the appropriate fire extinguisher for various classes of fire
7. Be able to locate, identify and shut-off gas, oxygen and oil valves in the plant
8. Be able to transmit an accurate alarm by box and telephone to the company fire department and also to the community fire and police departments

Each objective is to be tested at the end of trainee-text interaction using paper-and-pencil and possibly performance tests administered by fire officers.

16.5.1 → 16.6 The step format should have these characteristics:
1. Four steps per page
2. Page segment vertical design
3. Stimulus-response-feedback within the rectangular area based on 14.0 → 16.6, primarily:

    14.1.1 → 16.6   activity
    14.1.2 → 16.6   repetition
    14.1.3 → 16.6   reinforcement
    14.1.6 → 16.6   knowledge of results
    14.2.2 → 16.6   parts—whole relationship

14.3.5 → 16.6   anxiety level

14.3.6 → 16.6   motive and value organization

This will be input later to 4.3.3.

3.3 → 4.1.1 Training specialists select only those elements from 2.1 which are in the *DB* cell:

1. *D*; directly related to fire fighting and attitudes towards property
2. *B*; in the Introductory Course (IC)
3. *B*; in the Team Course (TC)
4. *B*; in the Team Refresher Course (TRC)

Time will not permit any instruction in *I* or *G* elements. The course structure allows only for *B* or Basic courses; Intermediate or Advanced courses are unauthorized.

4.1.1 → 4.1.2 Selection of pattern performed by training specialists:

1. IC
    a. Programmed text; Gestalt
    b. Classroom and fire ground, Non-Gestalt
2. TC—fire ground instruction; Non-Gestalt
3. TRC—fire ground instruction; Semi-Gestalt

4.1.2 → 4.1.3 Training specialists consulting with Fire Training Advisory Committee produce COTP:

1. Specific terminal behavioral objectives:
    a. IC programmed text (identified earlier)
    b. IC classroom
        1. Be able to identify each type of extinguisher for various classes of fire in five seconds per class
        2. Be able to tie one particular knot in five seconds
        3. Be able to specify three parts of the fire triangle in ten seconds
    c. IC fire ground
        1. Be able to spot a real fire and identify class in ten seconds
        2. Be able to select proper extinguisher from among five types in five seconds

3. Be able to completely extinguish each of three different classes of fire using hand extinguishers in 30 seconds
4. Be able to lay a 50-foot length of 1 1/2-inch fire hose; turn the valve on; select mist, spray or solid stream depending on class of fire; extinguish fire unassisted in 100 seconds
5. Be able to raise and climb a collapsed 16-foot ladder to a one-story building window unassisted in 25 seconds
   d. TC fire ground
1. Be able to lay 200-foot length of 2 1/2-inch fire hose; turn the valve on; select spray or solid stream depending on class of fire; use a three-man team; extinguish large fire in 240 seconds.
2. Be able to prime portable turbine pump using pond as a source of water; pump through two 1 1/2-inch lines; use a three-man team; extinguish two separate fires in 300 seconds. The pond may be frozen in New York, Kansas and Tennessee. Time extension to penetrate frozen water source using hand tools is 120 seconds.
   e. TRC fire ground is the same as TC fire ground.

4.1.3 → 4.3.1 Training specialists generate teaching points for each course and feed back 4.3.1 → 4.1.3 until COTP is completed.

Teaching points for human-instruction (classroom and fire ground) are combined with particular methods and placed in lesson plan format. Teaching points for programmed text are further processed 4.3.1 → 4.3.2 and will be explained shortly.

4.1.3 → 4.2.1 Training specialists create tests:
1. IC programmed text
   a. Paper and pencil post-test

    b. Performance post-test on transmitting an accurate alarm by box and telephone, and locating and shutting valves in buildings

2. IC fire ground—performance post-test on extinguishing one fire using hand extinguisher in 30 seconds.
3. TC fire ground—Performance post-test on three-man team extinguishing one fire using 2 1/2-inch line in 120 seconds
4. TRC fire ground is the same as TC fireground.

4.2.1 → 4.5.1 Training specialists conclude pre-tests are unnecessary since it is assumed at the outset that employees are unable to extinguish fires and require instruction. Therefore, 4.2.2 is bypassed. The tests will be used to validate all three courses in 4.5.1.

4.3.1 → 4.3.2 Training specialists screen teaching points through cubical matrix:

1. Objects: extinguishers, ladders, hose, hose tips, valves, rope, etc.
2. Actions: step-by-step procedures for tying knot, raising ladder, laying hose, etc.
3. Information: all other data not objects or actions.

4.3.2 → 4.3.1 Teaching points still obscure to training specialists regarding behavior to be elicited from trainee are fed back to 4.3.1 for rewriting. The output of 4.3.1 is always under *feedback* control.

4.3.2 → 4.3.3 Teaching points clearly stating the behavior are sent to 4.3.3 where each point is broken down into one or more steps. The steps are:

1. Introductory (I)
2. Generative (G)
3. Regenerative (R)
4. Synthetic (S)
5. Criterion (C)
6. Review (Re)

These appear in Figure 3-12 and were defined earlier.

16.6 → 4.3.3 About 120 steps are written; the step format previously established in 16.6 by training specialists is followed.

4.3.3 → 4.3.1 Steps must elicit teaching point behavior. Training
specialists find several points which are ambiguous, unclear
and poorly sequenced. These are reconstituted in 4.3.1 and
new steps are written in 4.3.3. The feedback loop 4.3.3 →
4.3.1 controls the output of 4.3.1 along with the 4.3.2 → 4.3.1
feedback loop.

3.3 → 18.1.1.2 Artwork consisting of drawings and a few halftones for the IC programmed text is produced. Art for the
loose-leaf cover is prepared. The classroom and fire ground
components of IC and TC and TRC do not require any software.

18.1.1.2 → 18.1.2.1 IC text is typed and edited; art is inserted.

18.1.2.1 → 18.1.2.2 Text is reproduced by offset press in the New
York plant's duplicating section (300 are run).

18.1.2.2 → 18.1.2.3 Thirteen hundred 3-hole loose-leaf covers and
masks are specified and purchased out-house; delivery is to the
New York plant. Three hundred 15-page texts are collated,
punched and stapled.

18.1.2.3 → 18.1.3.1 Thirteen hundred covers are received, inspected and repacked in lots of 250; 300 are to be used in the
tryout. Three hundred texts are packaged.

18.1.3.1 → 18.1.3.2 Three hundred spare covers and 1000 regular
covers are stored in the New York plant; texts are also stored
in the New York plant.

18.1.3.2 → 18.1.3.3 Covers and texts are placed under inventory
control; all covers are in stock, but 20,000 final texts have not
been printed.

18.1.3.3 → 18.1.3.4 → 4.4.1 Issue 300 spare covers and 300 texts
for tryout.

4.3.3 → 4.4.1 There are no ancillary materials, therefore 4.8 is
bypassed. The programmed text for IC is ready and has been
issued to training specialists for tryout.

4.1.3 → 4.3.1⎫ Lesson plans at the teaching point level to be used
15.0 → 4.3.1⎭ by fire officers, 15.0, are prepared by training

specialists under technical guidance of the Fire Training Advisory Committee. These combine each specific teaching point with a particular method for instructing it and set both down in a lesson plan format. Lesson plans are prepared for:

1. IC classroom and fire ground; two hours; used at company fire station
2. TC fire ground; two hours; used at company fire station

TRC lesson plans are not prepared at this time. They will be essentially the same as TC but may have to be updated when the new turbine pumps are introduced.

4.3 → 4.4.1 Lesson plans for use by 15.0 are now ready for try-out.

18.1.2.1 → 18.2.1 While software is being prepared, information directly from the 15.0 lesson plans provides a list of hardware items, some of them consumables, which are to be ordered. These consist of fire extinguisher refills and recharges, turbine pumps, ladders, 1 1/2-inch and 2 1/2-inch hose lines, nozzles, tips, protective clothing, etc. These are centrally purchased by the New York plant for delivery to California, Kansas, New York and Tennessee. Each plant is authorized to procure locally gasoline, oil, typical factory chemicals and wood for creating fire-ground fires used to train employees.

14.0 → 17.1 Training specialists and Fire Training Advisory Committee agree that a fire ground should be constructed at each plant. The purpose is to have real fires occur in a simulated and controlled setting so that employees can extinguish fires under realistic environmental conditions. The primary psychological bases for IC and TC are:

14.1.1 → 17.1	activity
14.1.2 → 17.1	repetition
14.1.3 → 17.1	reinforcement
14.1.4 → 17.1	generalization
14.1.5 → 17.1	drive in IC
14.1.6 → 17.1	knowledge of results
14.3.1 → 17.1	abilities

14.3.5 → 17.1   anxiety level

14.3.6 → 17.1   motive and value organization

14.3.7 → 17.1   group atmosphere

14.0 → 17.1 → 17.2 A human engineering consultant is engaged to assist training specialists and the Fire Training Advisory Committee in designing a simple, standard fire ground. His contribution is analyzing employee–fire equipment relationships as they exist in four plants (17.1) and designing a simulated building where fires can be fought internally and externally (17.2). The fire officers are better qualified to specify building details and to concentrate on fire safety aspects. Maintenance standards (17.3) are bypassed.

Tryout (4.4.1) is scheduled at the New York plant *without* waiting for fire ground construction to begin. The county's fire ground is used on a rental basis for tryout of the fire ground component.

17.1 → 17.2 → 19.1 A continually reusable building which cannot combust, located on a suitable plot of company land, is to be designed as a fire ground. Fires set inside should be extinguishable internally with hand extinguishers and externally with 1 1/2-inch and 2 1/2-inch hose lines layed from a typical plant standpipe or hydrant. Lesson plans (4.3) written by training specialists guided technically by the Fire Training Advisory Committee form the basis for design.

19.1 → 19.2 A fire ground consisting of a one-story welded steel building is constructed on large troughs on the edges of company property in California, Kansas, New York and Tennessee. Internally the building consists of 3200 square feet divided into a number of rooms containing obsolete machine tools, steel benches, etc. Everything is of noncombustible material. Design is by the Fire Training Advisory Committee who are assisted by plant public works engineers. Construction is by the plant's public works department. The Florida fire ground is to be designed later after experience in other plants has been obtained.

19.2 → 19.3 The plant public works department superintendent in each state turns over facility to plant fire chief.

19.3 → 5.3 Fire chiefs assign responsibility for fireground to the fire training officer.

18.2.1 → 18.2.2 ⎫ Hardware has been purchased centrally and is
18.2.2 → 18.2.1 ⎭ delivered to each plant. A maintenance schedule is established by the fire training officer in each plant. This includes a feedback, 18.2.2 → 18.2.1, to control the procurement of additional materials and equipment in advance of actual use.

18.2.2 → 18.2.2 ⎫ Materials and equipment are stored and a suffi-
18.2.3 → 18.2.2 ⎭ cient number prepared for issue during the 300-trainee tryout at the New York plant. Damaged equipment is locally repaired (18.2.3 → 18.2.2) either by fire officers or plant maintenance department. Factory level maintenance is performed by the original manufacturer. Thus, feedback controls the threshold readiness of hardware.

18.2.3 → 4.4.1 Materials and equipment for the 300-trainee tryout are at the New York plant.

4.2.1 → 4.4.1 Post-tests created originally in 4.2.1, based on 4.1.3 behavioral objectives, enter tryout.

13.3 Supervisors in the New York plant are requested to assign 300 trainees from the 5000-employee work force to the tryout for IC and TC. Supervisors follow original plans and men and women are assigned in the X-1 to X-16 paygrade range. Some training specialists suggested that a two-stage tryout be developed: first, 100 employees; then, after improving the courses, 200 employees. However, senior management countered with the suggestion that the New York plant, which had suffered a $3,000,000 loss needed *immediate* fire security. By conducting a one-stage tryout, 300 employees would be rapidly trained and this encouraged management to support the program more quickly. The post-tryout analysis (4.4.2) would be more rigorous under these circumstances.

4.4.1 Tryout is conducted. The IC programmed texts are issued individually and employees work through and turn them in. A test is administered when the materials are returned. The classroom and fire ground phase of IC is conducted with 50 classes of six persons. When all IC is completed, teams are identified and TC is conducted. Performance tests are administered at the end of each course.

4.4.1 → 4.4.2 Training specialists and Fire Training Advisory Committee analyze test data obtained from tryouts.

4.4.2 → 4.5.1 The analysis of results is compared with theoretical pre-test scores of zero and with performance criteria developed in 2.1 and 4.1.3. A number of problems are disclosed but these are not of a serious nature.

4.5.1 → 4.5.2 Training specialists and the Fire Training Advisory Committee accept the courses conditionally:

1. Several sections of programmed text need improving.
2. Cost per trainee for equipment and materials should be increased from $2.00 to $2.35.
3. Protective clothing (rubber coats, cap and boots) used by trainees at fire ground must be sanitized; this was not considered in original planning.
4. During TC, the second fire company cannot perform inspections but must be on standby at fire ground in case the actual fire gets out of control and endangers the premises; this was not considered in original planning.
5. Trainees who engaged in horseplay at the fire ground should be matched with those who knowingly spectated at fire disasters; this information should be processed into supervision and management channels for corrective action.
6. Information and fire ground training on portable turbine pumps should be deferred unless these pumps are actually installed throughout the plant.
7. Women employees represent a minority in the plants and the ladder raising and climbing instruction should be modified or made optional with them because during a disaster

it is unlikely they would have to raise ladders although they might have to climb down one previously raised.

8. When training is conducted during cold weather periods, the performance test times should be increased 30% to allow for snow and ice conditions at the fire ground.

9. Cold weather should not result in course cancellations since real disasters occur under all weather circumstances.

10. Several fire officers tended to wander from the lesson plan while instructing and this resulted in less time for essential trainee activities.

$4.5.2 \rightarrow 3.4 \rightarrow 3.5$
$4.5.2 \rightarrow 5.0 \rightarrow 6.0$ $\left.\right\}$ *Feedback to control the program:*
$4.5.2 \rightarrow 5.0 \rightarrow 13.0$

1. $3.5 \rightarrow 3.3 \rightarrow 4.3.3$: Rewrite 34 steps in several sections to improve programmed text; just rewriting steps does not remedy the situation and it is necessary to go to 4.3.1 and rewrite four teaching points. Feedback controls 3.4, 3.3, 4.3.3.

2. In one instance, adjusting step 4.3.3 through machine screw caused the limit switch to energize, and the training specialist reexamined 2.1 to find more specific behavior. The path traced is $4.3.3 \rightarrow$ screw $\rightarrow 2.1 \rightarrow 3.1 \rightarrow 3.2 \rightarrow 3.3 \rightarrow 4.1.1 \rightarrow 4.1.2 \rightarrow 4.1.3 \rightarrow 4.3.1 \rightarrow 4.3.2 \rightarrow 4.3.3$. Feedback controls 2.1.

3. In another instance, rewriting step 4.3.3 produced an error signal and the training specialist reviewed 4.1.3 to find more specific behavior. The path traced is $4.3.3 \rightarrow 4.6 \rightarrow 4.1.1 \rightarrow 4.1.2 \rightarrow 4.1.3 \rightarrow 4.3.1 \rightarrow 4.3.2 \rightarrow 4.3.3$. Feedback controls 4.1.1.

4. $3.4 \rightarrow 3.2$ Model does *not* allow this path; the recommendation to increase the cost per trainee from $2.00 to $2.35 is infeasible in this model. Therefore, corporate management cannot be approached for additional funds during the first year. However, the training organization has a contingency fund and authorizes transfer of $6000 from a special account without further approval. That is a feature

built into the finance system of the company to make projects run smoothly under local management control. Feedback controls 3.4.

5. 3.5 → 3.3 → 18.1.1 → 18.1.2.1 → 18.2.1 → 18.2.2: Fire training officer sets up protective clothing sanitizer at fire station and rotates all clothing through it to avoid personal health problems created by unsanitary gear. Feedback controls 3.4 and 3.3.

6. 3.4 → 3.2: Model does *not* allow this path; the recommendation to have second fire company cease inspections and remain on standby at the fire ground would seriously impair the fire security of the plant. However, all fire apparatus is two-way radio equipped, with the control center unit located at the telephone operator's console. Spare walkie-talkies on the alarm frequency are issued to each fire ground instructor. These are attached to the belt. When a need for assistance arises, any instructor merely presses a button on his unit and a siren is set off on the fire apparatus parked outside the building being inspected. Fire officers immediately rejoin the apparatus and speed directly to the fire ground. Elapsed time is in the order of five minutes or less. Feedback controls 3.4.

7. 4.5.2 → 5.1 → 5.2 → 5.1 → 6.1 → 6.4.3 → 6.4.1 → 6.4.3: Fire officers rate each trainee on proficiency at end of IC, TC and TRC. Normally these ratings are sent to the plant training organization where a record is made in the personnel files at the plant level. Employees who do not meet training standards for any reason are rated and reported to first-line supervision for remedial action. Feedback controls 5.1 and 6.4.3.

8. 3.5 → 3.3 → 4.1.1 → 4.1.2 → 4.1.3 → 4.3.1: Lesson plan content referring to portable turbine pumps is replaced with content dealing only with equipment deployed and procedures currently used in the particular plant. This policy is followed even if the fire ground is equipped with

the pumps while the plant has not yet had them installed in various buildings. A change-over in training occurs when the pumps are installed. Instruction then incorporates turbine pumping procedures. Feedback controls 3.4, 3.3 and 4.1.3.

9. $3.5 \rightarrow 3.3 \rightarrow 4.1.1 \rightarrow 4.1.2 \rightarrow 4.1.3 \rightarrow 4.3.1$: The terminal behavioral objective, "be able to raise and climb a collapsed 16-foot ladder to a one-story building window," stated in 4.1.3, is restated for female trainees *at their option* to "be able to climb a 16-foot ladder from a one-story building window." Feedback controls 3.4, 3.3 and 4.1.3.

10. $3.5 \rightarrow 3.3 \rightarrow 4.1.1 \rightarrow 4.1.2 \rightarrow 4.1.3 \rightarrow 4.2.1$: The IC fire ground performance test minimum times are *increased* when local weather is 30° F or lower, or there is reduced visibility due to snow or fog, or ground conditions are icy or snowbanked:
    a. Spot fire—10 to 13 seconds
    b. Select extinguisher—5 to 7 seconds
    c. Extinguish with hand device—30 to 39 seconds
    d. Lay 1 1/2-inch hose—100 to 130 seconds
    e. Raise and climb ladder—25 to 32 seconds
    The TC fire ground requirements are similarly increased:
    a. Lay 2 1/2-inch hose—240 to 312 seconds
    b. Use turbine pump; previously specified as 300 to 420 seconds. Feedback controls 3.4, 3.3 and 4.1.3.

11. $4.5.2 \rightarrow 4.7 \rightarrow 5.3 \rightarrow 5.1 \rightarrow 5.2 \rightarrow 13.2 \rightarrow 13.3$: Course cancellations normally occur only when fire officers are unable to instruct due to emergency assignments. However, if weather prevents employees from getting to the plant, the IC trainees may have man for man substitutions, but team assignments to a TC course may be cancelled and different teams substituted. Feedback controls 13.2.

12. $4.5.2 \rightarrow 5.1 \rightarrow 5.2 \rightarrow 5.3$: Every fire officer has been trained in ITC to instruct. Those actually assigned are subject to reassignment if they are unpopular with employees.

This is quickly detected by the fire chief or fire training officer during observation in 5.1. Comment by employees, (5.1 → 24.0 → 4.6 → 2.1 → 3.1 → 3.2 → 3.3 → 15.0) reaches the training specialists who are in constant communication through the Fire Advisory Training Committee with the appropriate fire training officer. Feedback controls 5.3, 4.6 and 2.1.

4.5.2 → 4.7 → 5.3 → 5.1 When all the changes resulting from the feedback are implemented, the *Fire Training Officer's Guide* is produced in 4.7 under sponsorship of the Fire Training Advisory Committee. Thus, all four operating plants, and Florida when opened, will have relatively uniform programs. Fire training officers manage instruction (5.3) using this guide. Instruction is conducted in 5.1.

4.5.2 → 3.4 → 3.5 → 3.3 → 18.1.1.2 → 18.1.2.1 → 18.1.2.2 → 18.1.2.3 → 18.1.3.1 → 18.1.3.2 → 18.1.3.3 Following artwork and typing changes, the programmed text is printed in a final first edition of 20,000 copies by the New York plant's duplicating section and issued to 5.1 on request. Inventory is controlled by scheduling in 13.3 → 18.1.3.3. In 5.1, trainees may retain the text after use. The 300 copies used in tryout are sent to employees via company mail so that each employee will have received his personalized programmed text.

13.3 → 5.1 Trainees are scheduled into IC and TC as prescribed in the CPB. Each plant is to conduct 825 IC classes of six trainees each, totaling 825 x 4 or 3300 classes company-wide. The number of trainees is 3300 x 6 or 19,800. Since 300 were trained during tryout the total is 19,800 + 300 or 20,100. Each plant also is to conduct 1650 TC classes. The TC class has three team members.

15.0 → 5.1 Fire officers conduct IC and TC. Illness, cancellations, transfers, retirements, new hires and similar perturbations require 3400 IC courses and 1702 TC courses by the end of year one.

5.3 → 5.1 Fire training officers manage IC and TC.

5.1 → 5.2 → 5.1 While IC and TC are being conducted, post-tests are used to evaluate employee proficiency. Evaluation (5.2) is performed by the fire training officer. Minor alterations in lesson plans and in equipment procedures are made periodically by fire officers. Feedback controls 5.1.

5.1 → 5.2 → 5.3 → 5.1 Alterations in plans and procedures considered important are made only by authorization of the cognizant fire training officer in each plant. Evaluation (5.2) is performed by the fire chief, training specialists and the fire training officer. Feedback controls 5.3.

5.1 → 5.2 → 5.3 → 19.2 Alterations at the fire ground which involve physically modifying the facility are made only by authorization of the fire training officer and are supported by funds from his budget. Evaluation (5.2) is performed by the fire chief, training specialists and the fire training officer. Feedback controls 5.3 and 19.2.

5.1 → 5.2 → 4.6 Alterations in lesson plans and programmed text necessitated by major changes in fire-fighting procedures and equipment are made (installation of automatic foam systems):
1. 4.6 → 2.1: Reanalyze the job and reconstitute the JA-HAA. Feedback controls 4.6 and 2.1.
2. 4.6 → 2.1 → 3.1: Reanalyze the parameters and reconstitute them. Feedback controls 4.6 and 2.1.
3. 4.6 → 2.1 → 3.1 → 3.2: Reanalyze parameter quantification and reassign values. Feedback controls 4.6 and 2.1.
4. 4.6 → 2.1 → 3.1 → 3.2 → 3.3: Reanalyze methods and introduce newly developed techniques. Feedback controls 4.6 and 2.1.
5. 4.6 → 4.1.1 → . . . 4.3.1: Make major changes in lesson plans. Feedback controls 4.6, 4.1.1 and 4.1.3.
6. 4.6 → 4.1.1 → . . . 4.2.1: Make major changes in tests. Feedback controls 4.6 and 4.1.1

6.4.3 → 6.4.1 → 4.6 In the event of a fire, those on the scene are primarily employees, their supervisors and fire officers. When the fire has been extinguished, industrial engineers study causes and review actions taken by all those present. Some of the information is gathered to satisfy insurance claims and to prevent the same type of fire reoccurring from the engineering viewpoint. Senior management has charged the industrial engineering department with responsibility for evaluating the effectiveness of the IC, TC and TRC with the plant fire departments. After 14 fires in California, Kansas, New York and Tennessee, the analyses based on comparison with previous fire experience reveal:

1. Most fires result from process breakdown and not employee carelessness.
2. Box and telephone alarms are transmitted within one minute after discovery of a blaze.
3. Employees in every instance are able to suppress the fire before the plant fire apparatus reaches the scene.
4. Evacuation of employees down ladders is accomplished without difficulty.
5. Fires are stopped in one-third the time.
6. Water damage causes only 1% of the loss.
7. Employee attitude at fires is serious and cooperative.

There is no need to utilize feedback to control 4.6.

The Value of Simulation

We have had an opportunity to run the Employee Fire Fighting problem through the model in the supplementary illustration with the objective of simulating to solve the problem. Simulation:

1. Caused the "player" to face up to specific procedures or functions represented by descriptors inside rectangles. The player either rationalizes an action or justifies bypassing the function. He cannot avoid decision-making.
2. Requires the player to consider a multitude of inputs to a single function, thus expecting more complex decision-making within the function.

3. Presents the player with alternatives and expects him to trade-off and be committed to one (hopefully, the best alternative) for the duration of that particular simulation.
4. Allows the player to modify his decisions by following feedback loops.
5. Permits the player to work out the problem on the paper model without incurring any expense or committing errors in real life.
6. Encourages quantitative thinking by making it relatively simple to construct numerical values and insert them throughout the problem.
7. Invites the player to invent different solutions to the same problem thereby producing *synthesis* in individuals who tend to be more comfortable with *analysis*.

Time permitting, the reader may wish to alter some of the parameters in the same problem and run them through the model to see what changes, if any, might appear. A different problem run through the same model would also be a worthwhile activity.

In this chapter, we have described a realistic training problem in a practical setting. A flowchart model of 164 subsystems designed to process a wide variety of situations was used to examine this problem. Simulation required the use of only 97 of the 164 subsystems. Various conditions, states, alternatives and possibilities were talked through and decisions were made. The quantification level was low—limited to a few simple calculations dealing with trainee population, instructor staffing, materials development cost and man-hours involved. Simulation occurs on paper and represents a low-cost, minimum effort activity to assist the training specialist in making decisions without ignoring any important subsystem.

5- the future
of training systems

Talking Through or Narration Simulation

The development of a training program can be a complex undertaking. Modeling and simulation reveal just how complex the decision-making can be. Many training specialists make equally complex decisions every day without resorting to a model and talking through several simulations. This is accomplished successfully only because of the breadth of experience which has produced, in effect, a similar decision-making network stored in the brain. For these training specialists, simulation is *covert* activity mentally performed at high speed. But what about less experienced training specialists, many isolated from the gentle guidance of senior training management?

Narration simulation is an *overt* activity which can substitute for experience. This is not to say that cerebration ceases when narration begins. Simply putting down and processing signals on paper forces the player to think in greater detail and in broader scope to solve a complex problem. Narration simulation is a game in which many players can participate together. In fact, solutions are enhanced when several players interact and challenge one another by insidiously contributing pitfalls, extraneous facts, cul-de-sacs and similar imaginative inventions. Sometimes the model has to be modified, but always decisions are made and a final solution results.

Narration simulation is a giant step towards quantification.

Computing with Pencil and Paper

One of the most convenient parameters to calculate is *time.* How long should a particular activity take from start to finish? Instruction time can be expressed in minutes; development time can be estimated in man-hours; travel or transit time can be computed in various appropriate units of minutes to walk, hours to fly, weeks to ship and deliver. It is possible to run the Employee Fire Fighting problem in Chapter 4 through again and just count the number of minutes and hours needed to accomplish each event. The enormity of a project first adjudged "simple" is quickly revealed through simulation! The author once studied an instruction system proposal and evaluated it using elapsed time as the basic unit of measurement (39). The planners had ignored time in their proposal, and the author likened their plan, *argumentum ad absurdum,* to a military situation in which the "mission must be *planned* and *completed* 14 minutes *before* the targets are identified. . . the pilot is so concerned with manipulating the missiles on his aircraft, he takes off from a carrier and flies his weapons to the target *without* the aircraft!" (39).

Another convenient parameter to calculate is *cost.* How much should a particular activity cost from start to finish? The unit can be U.S. dollars, Belgian francs or whatever. There are direct costs, which are very obvious and often appear as line items in a budget, indirect costs, which tend to be less obvious and are sometimes hidden away or manipulated as overhead expenses, and profit, which involves work performed out-house. Some costs always escape analysis and attempting to determine them is a foolish academic exercise. For example, if 20,000 programmed texts are to be stored in a room somewhere in the facility, does the storage area cost per month really matter if an invoice is never issued for the expense? It only matters when there is a clearcut charge against an account and money changes hands, either in fact or on paper. Generally, all practical costs can be computed just as time is measurable—during narrative simulation and before actual implementation of a training program. If costs are to be calculated, then time estimates should be determined first. Costs derived from time estimates tend to be more realistic and closer to the final, total expense after implementation. Again, it is possible to run the Employee Fire Fighting problem through the model and just add dollars.

A less convenient parameter to calculate is *quality*. A better term is *effectiveness* but it has, by implication, an action-oriented meaning. Quality and effectiveness are not subject to mechanistic measurement using clocks and adding machines as are time and cost. Effectiveness is gradually yielding to quantification in the training field through the use of judges (experts) who are now expressing their measurement criteria on paper. The criteria can be applied by less-experienced persons, since they are measurable and achievable. Ryan's SPAMO Quality Test considers *s*pecificity, *p*ertinence, *a*ttainability, *m*easurability and *o*bservability as requirements in dealing with behavioral objectives (40).

Head count, the *number* of training program graduates, is often used either to predict or to count after the program ends, as an overall appraisal of system *capacity* or *efficiency*. The dropout rate may be important to some who view the system as a petroleum fractionation column where various products are distilled off at different times during the processing.

The basic concept in creating a quantifiable relationship is delineated in Figure 5-1. In 1.0, the term *a* is defined—not the value of *a* but the meaning of *a*. The term *b* is defined in 2.0. These are input to 3.0 and multiplicative relationship establishes this new relationship as *c*. If one wishes to have an output from 6.0, what might it be? It would always be the numerical value of *c* which results from the events occurring inside 6.0. $C = ab$ is an algebraic equation. Terms *a* and *b* are variables and so long as

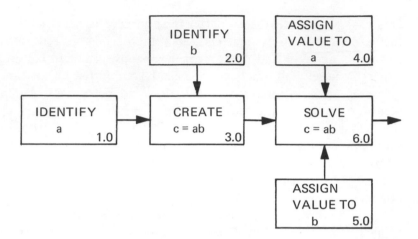

Figure 5-1.   Creating a quantifiable relationship.

numerical values are assigned to *a* and *b* and then input to 6.0, there will always be an output *c*. Each descriptor inside the rectangle is an expression of the real-life usefulness of an object, action or information, precisely and unambiguously. The rectangle, descriptor and point-numeric code together represent a *function*; see Figure 1-8, Chapter 1.

Figure 5-1 is a little *flowchart model,* as you might suspect! Mathematicians and engineers call *c = ab* a *mathematical model.* The contents of 6.0 are, by definition, a mathematical model.

Theoretically, it should be possible to take a large, complex flowchart model such as the one in the supplementary illustration and replace each descriptor with one or more algebraic equations of the *c = ab* variety. The result would be a flowchart model completely mathematized so that all descriptors are represented by mathematical relationships or equations. Figure 5-*2a* depicts the simplified flowchart model of an automotive mechanic training system in operation using LOGOS and produced in a course conducted by the author. Figure 5-*2b* is the mathematized flowchart model of the 5-*2a* analog prepared by Brooks (18).

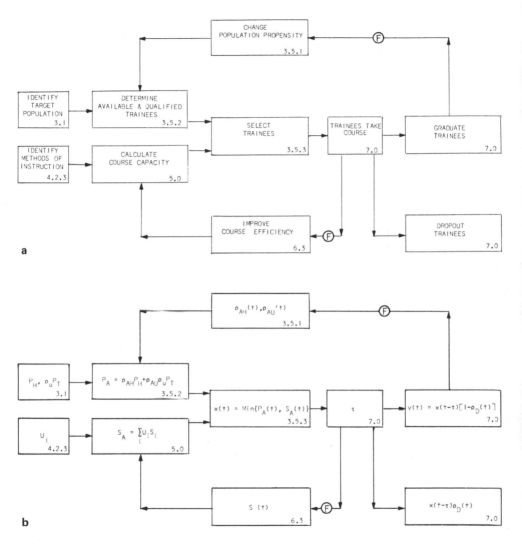

Figure 5-2. a. *Flowchart model of automotive mechanic training system in operation.* b. *Mathematized model of Figure 5-2a.*

Computing with a Digital Computer

Once it is possible to produce a mathematized flowchart model similar to Figure 5-2*b*, it shouldn't be necessary to rely solely on clocks and adding machines. The power of the present-day computer makes it a fairly routine matter for a computer programmer to convert an equation into a series of statements using a language such as BASIC, FORTRAN or PL/1 which the computer will understand.

The output equations appearing in Figure 5-2*b* have been written in Fortran II and these are listed in Figure 5-3 as a series of statements.

When the program in Figure 5-3 is read in and executed, it is possible to converse with the computer and conduct a simulation. This results in an on-line printout of a graph having two curves, which represent population and capacity, plotted on axes marked "time in courses" and "graduates in thousands." Even so, the FORTRAN program in Figure 5-3 is not highly conversational and an improved version of the same program written in LYRIC appears in Figure 5-4. The simulation program is written using the FOR statements in LYRIC and results in the computer-assisted instruction program shown in Figure 5-5.

Figure 5-5 represents a fairly sophisticated, conversationally interactive program. After asking the training specialist for his name, the computer acknowledges his identity and requests him to enter the number of instructors, number of computers and the number of students per computer based on the Brooks models in Figure 5-2*a* and *b*. The training specialist enters 1000, 10 and 100. The computer next requests the course length and the plotting increment. The training specialist enters 0.5 and 1. The computer finally asks for the values of K and C and is given 10^{-6} (written 1E-6) and 10^{-4} (written 1E-4). At this moment, the computer cranks out the two curves in Figure 5-5 and thanks the training specialist for a fine job. Points * and $ are connected with lines by hand, since the teleprinter is unable to draw vectors. In a CRT console system, the same curves would be machine drawn.

```
PRINT 10
10 FØRMAT(SENTER NØ. ØF INSTRUCTØRS, CØMPUTERS, STUDENTS PER CØMPUTERS/)
ACCEPT *,TEACHERS,CØMPUTERS,STUDENTS
PRINT 20
20 FØRMAT(SENTER LENGTH ØF CØURSE, AND INCREMENT FØR PLØTTINGS/)
ACCEPT *,TIME,BTIME
PRINT 30
30 FØRMAT(SENTER K AND CS/)
ACCEPT *,AK,C
PRINT 40
40 FØRMAT(//////)
PRINT 50
50 FØRMAT(STIME IN CØURSESS,15X,SGRADUATES IN THØUSANDSS/)
PRINT 60
60 FØRMAT(25X,S5S,13X,S10S,13X,S15S,13X,S20S)
PA=-100
ATIME=0
CTIME=0
CØMSTU=CØMPUTERS*STUDENTS
DØ 70 I=1,100
TT=.9*(TEACHERS*30*(1-(2.718**(-(((ATIME-1)*TIME)/3))))+CØMSTU)
TTT=45E5*(AK*PA+C)
CALL PLØT(2,CTIME,TT,20000.,0.,TTT,20000.,0.,DUM4,DUM5,DUM6)
PA=TTT
ATIME=ATIME+BTIME
70 CTIME=ATIME*TIME
END
```

Figure 5-3. Listing of FORTRAN II simulation program.

```
100        CØM THIS ILLUSTRATES A MATHEMATICAL MØDEL SIMULATIØN
101        CØM WHICH IS CØPYRIGHT 1970 BY EDUCATIØN AND TRAINING
102        CØM CØNSULTANTS CØ. AND CANNØT BE USED WITHØUT PERMISSIØN.
103        PRØ KINDLY TYPE YØUR FULL NAME AND PRESS RETURN KEY
104        ANS
105        STØ NAME
106        SHØ THANK YØU, /NAME/, FØR ENRØLLING IN THIS LESSØN.
107        PRE NØW, WE WILL RUN A SIMULATIØN USING THE MATHEMATICAL
110        '   MØDEL DEVELØPED BY CARL N. BRØØKS IN THE JUNE, 1969,
115        '   ISSUE ØF EDUCATIØNAL TECHNØLØGY.
120        PRE ENTER NØ. INSTRUCTØRS,CØMPUTERS,STUDENTS PER CØMPUTER
125        FØR       ACCEPT *,TEACHERS,CØMPUTERS,STUDENTS
130        PRE ENTER LENGTH ØF CØURSE, AND INCREMENT FØR PLØTTING
135        FØR       ACCEPT *,TIME,BTIME
230
235        PRE
240        PRE THIS CØMPLETES THE DEMØNSTRATIØN IN WHICH YØU
245        '   WERE ABLE TØ ENTER VARIABLES AND SIMULATE ØN-LINE.
260        SHØ THANK YØU, /NAME/, FØR A FINE JØB.
265        PRE
270        END
```

Figure 5-4. LYRIC program with FORTRAN statements of simulated programs.

```
KINDLY TYPE YØUR FULL NAME AND PRESS RETURN KEY

? LEØNARD C. SILVERN
THANK YØU, LEØNARD C. SILVERN, FØR ENRØLLING IN THIS LESSØN.

NØW, WE WILL RUN A SIMULATIØN USING THE MATHEMATICAL
MØDEL DEVELØPED BY CARL N. BRØØKS IN THE JUNE, 1969,
ISSUE ØF EDUCATIØNAL TECHNØLØGY.

ENTER NØ. INSTRUCTØRS,CØMPUTERS,STUDENTS PER CØMPUTER
1000,10,100

ENTER LENGTH ØF CØURSE, AND INCREMENT FØR PLØTTING
0.5,1

ENTER K AND C
1E-6,1E-4
```

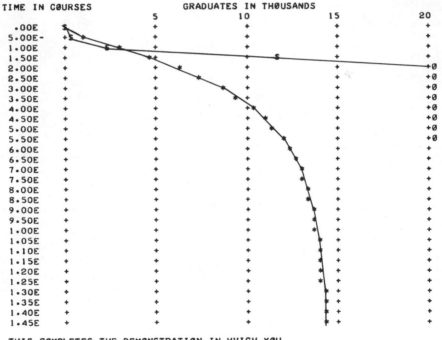

```
THIS CØMPLETES THE DEMØNSTRATIØN IN WHICH YØU
WERE ABLE TØ ENTER VARIABLES AND SIMULATE ØN-LINE.
THANK YØU, LEØNARD C. SILVERN, FØR A FINE JØB.
```

Figure 5-5. Trainee-computer conversational interaction based on standard LYRIC strategy and coding. (Points * and $ were connected with lines by hand.)

We have briefly described the present state of the art, working step-by-step, to:

1. Create a quantifiable relationship (Figure 5-1)
2. Convert a flowchart model (Figure 5-2*a*) to its mathematized counterpart (Figure 5-2*b*)
3. Write a FORTRAN computer program (Figure 5-3) or write a LYRIC computer program (Figure 5-4) representing the model which will permit simulation
4. Conduct the simulation (Figure 5-5) on-line with the computer

It is possible to change the input values and repeat the simulation iteratively until an acceptable set of curves (solutions) is generated. The example given is an extremely short simulation simply to illustrate the concept of quantification and implementation on a standard time-sharing computer. Without much imagination, it is possible to visualize this technique applied to a larger, complex model such as the flowchart in the supplementary figure (19). The narrative simulation could be completely replaced with computer simulation. The computer asks several questions and the training specialist keys in *either* numerical data or words in sentences. Based on this input, the computer prints out alternative decisions and asks for a selection decision. This continues until the problem is completely simulated on a conversationally interactive basis. Such parameters as time, cost and numbers of persons are easily handled but this is not yet true with quality and effectiveness. It is today within computer state of the art to have three or four training specialists each working on his own CRT console and operating on the same problem under simulation. They can interact with each other as well as with the problem being simulated. This aspect of time-sharing means the specialists could, for example, be at plants in California, Kansas, New York and Tennessee (sound familiar?) and run a computer simulation simultaneously without traveling to one particular location. The author referred to this as a *man-machine-man system* in 1963 (23). Now, nearly a decade later, the technique has been perfected.

It is a relatively routine matter to simulate problems on a model in the computer as complex´as that in the supplementary figure. Size or complexness, per se, are not obstacles. However, *this can only be accomplished if the model has been mathematized*—and there is where the future action is! Unless we can reduce words to quantitative relationships expressed as equations, the going will be hard. Lippitt brings his organizational insight to bear on this issue, which continues to be resisted in areas where attitudes are involved (41).

Training as a professional field has come a long way in only 50 years, since the early days of apprentice training in the railroad and shipbuilding industries. Tolerated by most, despised by some and abused by all, training can achieve startlingly new and exciting techniques and can become an equal in the corporate hierarchy. In this book, we have attempted to reinforce its traditionally shaky underpinings by pouring a new foundation of logical thought and systematic action through systems engineering.

references

1. Silvern, Leonard C. *Systems Engineering of Education 1: The Evolution of Systems Thinking in Education*, Education and Training Consultants Co. 1971.
2. Rau, John G. *Optimization and Probability in System Engineering*, New York: Van Nostrand Reinhold Co., 1970.
3. Chestnut, Harold *Systems Engineering Tools*, New York: John Wiley & Sons, 1965.
4. Silvern, Leonard C. "Systems Approach—What Is It?" *Educational Technology*, August 30, 1968.
5. _____ *Systems Engineering of Education VI: Principles of Computer-Assisted Instruction Systems*, Education and Training Consultants Co., May 1970.
6. _____ *Systems Engineering of Education XIII: Model For Producing Models* (slide/tape presentation), Education and Training Consultants Co., February 1971.
7. Shelley, Carl B. and Vaugh Groom. "The Apollo Flight Controller Training System Concept and its Educational Implications," *Computer-Assisted Instruction, Testing and Guidance*, 1970.
8. Porter, Arthur. *Cybernetics Simplified*, New York: Barnes & Noble, Inc. 1969.
9. DeGreene, Kenyon B. "Systems Psychology," in *Systems Psychology*, New York: McGraw-Hill Book Co., 1970.

10. Klir, Jiri and Miroslav Valach. *Cybernetic Modelling,* New York: D. Van Nostrand Co., 1967.

11. Silvern, Leonard C. "A Cybernetic System Model for Occupational Education," *Educational Technology,* January 30, 1968.

12. _____ "Cybernetics and Education K-12," *Audiovisual Instruction,* March 1968.

13. _____ "LOGOS, A System Language for Flowchart Modeling," *Educational Technology,* June 1969.

14. _____ *LOGOS Language for Flowchart Modeling* (slides/tape/text-workbook), Education and Training Consultants Co., 1970.

15. Goldberg, Rube. "Rube Goldberg Speaks Out on Design Simplicity," *EEE*; December, 1969.

16. Halsey, George D. *Training Employees,* New York: Harper & Bros., 1949.

17. Silvern, Leonard C. and Carl N. Brooks. *Systems Engineering of Education VIII: Quantitative Models for Occupational Teacher Utilization of Government-Published Information,* Education and Training Consultants Co., 1969.

18. Brooks, Carl N. "Training System Evaluation Using Mathematical Models," *Educational Technology,* June 1969.

19. Silvern, Leonard C. *Systems Engineering of Education V: Quantitative Concepts for Education Systems*, Education and Training Consultants Co., 1972.

20. "How to help Mail Service work better for you," Washington, D.C.: U.S. Government Printing Office, 0-425-246, 1971.

21. Silvern, Leonard C. *Fundamentals of Teaching Machine and Programmed Learning Systems* (a Programmed Instruction Course), Education and Training Consultants Co., 1964.

22. Forrester, Jay W. *Industrial Dynamics,* Cambridge: M.I.T. Press, 1961.

23. Silvern, Leonard C. "Shaping and Controlling Human Behaviour in Man-Machine Systems, *Proceedings of the Institution of Mechanical Engineers* (London) Vol. 177, No. 34, 1963.

24. *The Training Within Industry Report 1940-1945,* Bureau of Training, War Manpower Commission, Washington D.C.: U.S. Government Printing Office, September 1945.

25. Silvern, Leonard C. *Methods of Instruction,* Hughes Aircraft Co., 1957, 1962.

26. _____ *Basic Analysis* (a Programmed Instruction Course), Education and Training Consultants Co., 1965.

27. _____ "Change in the Training Director's Job," *Journal of the American Society of Training Directors,* February 1961.
28. Lange, Phil C., Editor. *Programmed Instruction,* 66th Yearbook, National Society for the Study of Education, 1967.
29. Sedlik, Jay M. *Systems Engineering of Education XIV: Systems Techniques for Pretesting Mediated Instructional Materials,* Education and Training Consultants Co., 1971.
30. Sippl, Charles J. *Computer Dictionary and Handbook,* Indianapolis, Ind.; H.W. Sams & Co., 1966.
31. Larer, Murray. "User's Influence on Computer Systems Design," *Datamation,* October 1969.
32. Rigney, Joseph W. "Maintainability: Psychological Factors in the Persistency and Consistency of Design," *Systems Psychology,* Kenyon B. DeGreene, Editor, New York: McGraw-Hill Book Co., 1970.
33. *Field Service and Support Exhibit "A,"* 59H- 4929/4722-500, Hughes Aircraft Company, Culver City, California, undated (about 1962).
34. Silvern, Leonard C. *Instructors Manual-State Rescue Training Program,* New York State Division of Safety, Albany, N.Y., 1951.
35. _____ *Fundamentals of Teaching Machine and Programmed Learning Systems Guide,* Education and Training Consultants Co., 1964.
36. Hilgard, Ernest R. "Learning Theory and It's Applications," *New Teaching Aids for the American Classroom,* Institute for Communication Research, Stanford University, 1960.
37. Knowles, Malcolm. *Learning Theory and Adults,* Houston: Gulf Publishing Co. (in press).
38. Patton, John A. "The First-Line Supervisor: Industry's Number One Problem," *Business Management,* September 1971.
39. Silvern, Leonard C. *Systems Engineering of Education IV: Systems Analysis and Synthesis Applied Quantitatively to Create an Instructional System,* Education and Training Consultants Co., 1969.
40. Ryan, T. Antoinette (Program Director). *Model of Adult Basic Education in Corrections,* Education Research & Development Center, University of Hawaii, Honolulu, April 30, 1970.
41. Lippitt, Gordon L. *Models for Change,* awaiting publication.

index

Forrester, J.W., 39
FORTRAN
    computer-assisted instruction,
        95-96
    simulation, 158-61
Functions
    abort, 85
    critical, 25-26
    modeling, 3, 13, 156
    reject, 85
    summer, 22-23, 34

G

Gain
    open-loop, 35-37
    pre-test/post-test, 85, 87, 116, 118
Goldberg, R., 26

H–I–J–K

Hilgard, E.R., 120-23
Instruction methods
    computer-assisted instruction
        (CAI), 60-62, 80, 111
    film, TV, 92
    human-instruction, 59, 79-83, 111,
        123
    machine-instruction, 59, 83-90, 111
    mix, 66
    multimedia, 90-91
    programmed instruction, 59-60,
        64, 79-80, 85-91
Job
    acquisition of information, 68-69
    analysis (JA-HAA), 50, 55-59,
        68-73
    content, 73
    criticality, 71
    definition, 55

Job (*cont.*)
    DIG relationship, 69, 117
    elements, 68
    equations, 68-69
    hierarchy of stages, 68, 73
    level of proficiency, 70, 117
    produce in-house, 86
    synthesis (JS-HAS), 125
    time available, 74
Knowles, M., 123

L

Language
    CAI, 60, 62
    definition, 13
    jargon, 1
    LOGOS, 12-38
    natural, 100
Level of detail
    applications, 38, 64
    code, sequencing, 27-29, 33
        structure, 16
    definition, 15
    generalized model, 8
    resolution, 15, 67
    Sedlik model, 92
Lippitt, G.L., 162
Logistics, logistical support, 50, 63,
        91, 102-05

M

Markov chain, 113
Modeling
    anasynthesis, 6
    changing model, 82
    LOGOS language, 12-38, 39,
        156-57